A Primer of Population Genetics

A PRIMER OF POPULATION GENETICS

Third Edition

Daniel L. Hartl
Harvard University

Sinauer Associates, Inc. • *Publishers*
Sunderland, Massachusetts • *U.S.A.*

The Cover

Genetic variation in color among individuals of the seastar *Pisaster ochraceus* on the Olympic Peninsula. This organism plays a large role in the food web of intertidal habitats throughout the Pacific Northwest and is often considered a keystone species. During high tides *P. ochraceus* feeds on barnacles, mussels, and many other prey. Its predatory habits and feeding strategies have made it a favorite subject of ecologists studying such topics as predator-prey interactions, competition, community structure, and life-cycle evolution. (Photograph © Darrell Gulin/Dembinsky Photo Associates.)

A PRIMER OF POPULATION GENETICS, 3RD EDITION

Copyright © 2000 by Sinauer Associates, Inc. All rights reserved. This book may not be reproduced in whole or in part for any purpose. For information address Sinauer Associates Inc., 23 Plumtree Road, Sunderland, MA 01375 U.S.A.

Fax: 413-549-1118
Email: publish@sinauer.com
www.sinauer.com

Library of Congress Cataloging-in-Publication Data

Hartl, Daniel L.
    A primer of population genetics/Daniel L. Hartl.--3rd ed.
        p. cm.
    Includes bibliographical references and index.
    ISBN 0-87893-304-2 (pbk.)
        1. Population genetics. 2. Quantiative genetics. I. Title.

QH455.H36 1999
576.5'8--dc21   99-045796

Printed in U.S.A.
5 4 3

This book is dedicated to the community of scholars and students who, over the years, have helped to create an exciting, productive, and mutually supportive research environment by giving generously of their time, energy, ideas, imagination, good humor, friendship, and hard work.

Stan Sawyer	Dan Dykhuizen	Elena Lozovskaya
Jim Ajioka	Lara Forde	Ray Miller
Zé Ayala	Dan Garza	Etsuko Moriyama
Isabel Beerman	Louis Green	Tendai Mutangadura
Douda Bensasson	Barry Hall	Javare Nagaraju
Susan Biel	Elisabeth Hauschteck	Yunsun Nam
Nathan Blow	David Haymer	Kaare Nielsen
Justin Blumenstiel	Chuck Hill	Dmitry Nurminsky
Fidelma Boyd	Ann Honeycutt	Eithne O'Brien
Glenn Bryan	Jim Jacobson	Howard Ochman
Carlos Bustamante	Bob Jones	John Parsch
John Campbell	Hans Jungen	Dmitri Petrov
Pierre Capy	Nizam Kettaneh	Steve Rich
John Carulli	Ben Kirkup	Nancy Scavarda
Duccio Cavalieri	Akihiko Koga	Jennifer Schutzman
Belinda Chang	Fedya Kondrashov	Mark Siegal
Diana Childress	Dan Krane	David Smoller
Andy Clark	Jeff Lawrence	Dave Sullivan
Olga Danilevskaya	Danne Lidholm	Courtney Timmons
Dan De Aguiar	Allan Lohe	Jeff Townsend
Tony Dean	Laurel Mapes	Patricia Tsai
Jean de Framond	Kyoko Maruyama	Cristina Vieira
Rob DeSalle	Meetha Medhora	Jorge Vieira
Bob DuBose	Colin Meiklejohn	Yaping Xu
Jack Dunne	Rossella Milano	Susan Yuknis

Contents

Chapter 4 The Genetic Architecture of Complex Traits 151

Preface

The field of population genetics has changed dramatically since I chose it as a career. The transformation came about from the abundance of experimental data generated by the use of molecular methods to study genetic polymorphisms. Present in virtually all organisms, molecular polymorphisms allow populations to be studied without regard to species or habitat, and without the need for controlled crosses, mutant genes, or for any prior genetic studies. It is for this reason that a familiarity with population genetics has become essential for any biologist whose work is at the population level. These fields include evolution, ecology, systematics, plant breeding, animal breeding, conservation and wildlife management, human genetics, and anthropology. Population genetics seeks to understand the causes of genetic differences within and among species, and molecular biology provides a rich repertoire of techniques for identifying these differences.

Organized as a "user's guide," this book is for undergraduate students, graduate students, and professionals in biology and other sciences who desire a concise but comprehensive introduction to population genetics. Chapter 1 introduces the basic concepts of molecular genetics and examines the principal methods by which DNA (or proteins) can be manipulated to reveal genetic polymorphisms in any population. Chapter 1 also includes the principles of population genetics underlying the organization of genetic variation in populations, with special emphasis on random mating, linkage equilibrium and disequilibrium, and inbreeding. Chapter 2 examines the evolutionary processes that can change allele frequencies, including mutation, migration and population admixture, natural selection of various types, and random genetic drift. This chapter also includes an elementary explanation of the diffusion equations as used in population genetics. Chapter 3 is the core of molecular population genetics. It includes the analysis of nucleotide polymorphism and diversity based on coalescents, patterns of change in nucleotide and amino acid sequences with special emphasis on codon usage bias and amino acid polymorphisms, inferences based on comparisons of levels of polymorphism and divergence, molecular phylogenetics, and the population dynamics of transposable elements. Chapter 4 focuses on com-

plex traits whose expression is influenced by multiple genes and environmental factors. It examines genetic effects on the components of phenotypic
variation and the correlations between relatives, the evolution of quantitative traits in natural populations, and comes full circle with the use of molecular polymorphisms and candidate genes in the identification of quantitative trait loci underlying complex inheritance.

Although the text introduces a significant number of equations, the emphasis is on explanation rather than derivation. Only elementary algebra is
necessary to follow most of the material, but a familiarity with basic calculus
is helpful for understanding diffusion equations, Poisson random fields, and
complex threshold traits. Mathematical symbols are used consistently
throughout the book; when results from the theoretical literature are quoted,
I have changed the original symbols as necessary to maintain consistency.

Modern population genetics makes liberal use of acronyms, some for different types of DNA polymorphisms (RAPD, AFLP, RFLP, SNP, VNTR,
SSLP) and others for theoretical concepts used in analyzing such polymorphisms (HWE, LD, RGD, HKA, PRF, QTL). These are all defined in the text,
and their interrelations and implications are discussed. For ease of reference,
a list of acronyms and other common abbreviations used in population
genetics immediately follows this preface. The text also includes numerous
practical examples showing how the theoretical concepts are applied to
actual data. To reinforce ideas, each chapter includes about 20 problems at
the end, arranged approximately in order of difficulty. To keep the book to
manageable size, the answers are not included. The solutions, worked out
in full, can be found in the genetics section of the publisher's website at
www.sinauer.com/hartl/html. I will be grateful for any comments or suggestions, especially pointing out errors or descriptions that are unclear. My
email address is dhartl@oeb.harvard.edu.

Acknowledgments

First and foremost I wish to thank my long-term collaborators and friends,
Stanley A. Sawyer, Daniel E. Dykhuizen, and Elena R. Lozovskaya. They
have coauthored over 70 research papers with me, and their ideas and support have contributed to many other papers in which they are not formally
acknowledged. Their kindness, generosity, and creativity have made my pro-
fessional life pleasant as well as productive. I am also grateful for the contin-
uing cheerfulness and hard work of the personnel in my laboratory, who
endured months of my preoccupation, vacant stares, and curt conversations
while I was struggling with the revision. These include Eithne O'Brien, Elena
Lozovskaya, Dmitry Nurminsky, John Parsch, Kaare Nielse, Dmitri Petrov,
Dan De Aguiar, Carlos Bustamante, Jeffrey Townsend, Colin Meiklejohn,
Justin Blumensteil, Susan Yuknis, Isabel Beerman, Yunsun Nam and Fedya

Kondrashov. My family endured the same annoying traits at home, as well as long hours with me upstairs working in my study, and I am grateful for their patience. Special thanks goes to Andrew G. Clark, coauthor of *Principles of Population Genetics*, Third Edition (Sinauer Associates, 1997). Thanks also to Roger Milkman, Trudy Mackay, Glenys Thomson, Kenneth Weiss and James Jacobson, who thoroughly reviewed the second edition of *A Primer of Population Genetics* and made many suggestions for improvement.

No book sees print without the skill and effort of publishing professionals, and the staff at Sinauer Associates is among the best. Thanks to editors Andy Sinauer and Carol Wigg and to the production team supervised by Christopher Small, with special thanks to Michele Ruschhaupt for her care and patience in creating the final pages, including the graphic art.

DANIEL L. HARTL
CAMBRIDGE, MASSACHUSETTS

Acronyms and Abbreviations

A	Adenine, the purine base, or a nucleotide containing adenine.
AFLP	Amplified fragment length polymorphism.
bp	Base pair, a unit of length in nucleic acids equal to 1 base pair.
C	Cytosine, the pyrimidine base, or a nucleotide containing cytosine.
CAI	Codon adaptation index, a measure of codon usage bias.
CAPS	Cleaved amplified polymorphic site.
χ^2/L	Scaled chi-square, a measure of codon usage bias.
Cov	Covariance.
cM	Centimorgan, a unit of length in a genetic map equal to 1% recombination.
cpDNA	Chloroplast DNA.
DNA	Deoxyribonucleic acid.
E	Expected value, the mean.
ENC	Effective number of codons, a measure of codon usage bias.
5′ end	The end of a nucleic acid strand containing a free 5′ phosphate group on the sugar.
G	Guanine, the purine base, or a nucleotide containing guanine.
Gb	Gigabase, a unit of length in nucleic acids equal to 10^9 bases or 10^9 base pairs.
GEA	Genotype-environment association.
GEI	Genotype-environment interaction.
GSI	Genotype-by-sex interaction.
HKA	Hudson-Kreitman-Aguadé test of polymorphism and divergence.
HWE	Hardy-Weinberg equilibrium (genotype frequencies p^2, $2pq$, q^2).

IBD	Identity by descent.
indel	Insertion or deletion.
IS	Insertion sequence, a type of transposable element.
kb	Kilobase, a unit of length in nucleic acids equal to 10^3 bases or 10^3 base pairs.
LD	Linkage disequilibrium.
LE	Linkage equilibrium.
lod or **LOD**	The log odds score; logarithm of the likelihood ratio.
Mb	Megabase, a unit of length in nucleic acids equal to 10^6 bases or 10^6 base pairs.
ML	Maximum likelihood.
MLE	Maximum likelihood estimate.
MRCA	Most recent common ancestor of two or more biological taxa or sequences.
mRNA	Messenger RNA.
mtDNA	Mitochondrial DNA.
MY or **Myr**	Millions of years.
MYA or **Mya**	Millions of years ago (sometimes abbreviated Ma).
PCR	Polymerase chain reaction.
Pr{X}	Probability of event X or outcome X.
PRF	Poisson random field.
QTL	Quantitative trait locus.
RAPD	Random amplified polymorphic DNA.
RFLP	Restriction fragment length polymorphism.
RGD	Random genetic drift.
RNA	Ribonucleic acid.
SCAR	Sequence-characterized amplified region.
SNP	Single-nucleotide polymorphism.
SPAR	Single-primer amplified region.
SSCP	Single-stranded conformational polymorphism.
SSLP	Simple sequence length polymorphism.
STR	Simple tandem repeat.
STRP	Simple tandem repeat polymorphism.
STS	Sequence-tagged site.
T	Thymine, the pyrimidine base, or a nucleotide containing thymine.
Tb	Terabase, a unit of length in nucleic acids equal to 10^{12} bases or 10^{12} base pairs.

TE	Transposable element.
3′ end	The end of a nucleic acid strand containing a free 3′ hydroxyl on the sugar.
U	Uracil, the pyrimidine base, or a nucleotide containing uracil.
Var	Variance.
VNTR	Variable number of tandem repeats polymorphism.
X	X chromosome.
Y	Y chromosome.
*	Statistical significance at the 5% probability level.
**	Statistical significance at the 1% probability level.

CHAPTER 1

Genetic Variation

Population genetics is more important today than ever before. One reason is that it has become possible to study genetics without focusing on mutant organisms that manifest visible differences, such as peas that are round or wrinkled, or fruit flies with red eyes or white eyes. Another reason is that it has become possible to study genetics without doing controlled crosses. The traditional requirements for mutants and controlled crosses have become dispensable because the discovery of genetic differences between organisms—even between species—is no longer limited to those differences that reveal themselves by the segregation of genes in pedigrees according to the time-honored principles of genetic inheritance first described by Gregor Mendel in 1866. In today's biology, genetic differences between organisms are often found by direct molecular analysis of DNA or proteins. The direct study of genes and gene products, without the need for crosses, means that detailed genetic analysis is no longer limited to domesticated animals, cultivated plants, and the relatively small number of experimental organisms that can be cultured in the laboratory. Genetic analysis is possible in any organism. It is for this reason that the concepts and experimental approaches of population genetics have come to pervade virtually every area of modern biology.

In its broadest sense, **population genetics** is the study of naturally occurring genetic differences among organisms. Genetic differences that are common among organisms of the same species are called genetic **polymorphisms,** whereas genetic differences that accumulate between species constitute genetic **divergence.** We may therefore define population genetics as the study of polymorphism and divergence.

GENETIC AND MOLECULAR ESSENTIALS

Although the genetic principles underlying population genetics are, for the most part, simple and straightforward, it may be helpful to preface the discussion with a few key definitions.

Genotype and Phenotype

Gene is a general term meaning, loosely, the physical entity transmitted from parent to offspring during the reproductive process that influences hereditary traits. The set of genes present in an individual constitutes its **genotype.** The physical or biochemical expression of the genotype is called the **phenotype.** There is a fundamental distinction between genotype and phenotype because, in general, there is not a one-to-one correspondence between genes and traits. Most complex traits, such as hair color, eye color, skin color, height, weight, behavior, life span, and reproductive fitness are influenced by many genes. Most traits are also influenced more or less strongly by environment. This means that the same genotype can result in different phenotypes, depending on the environmental contribution, and likewise that that same phenotype can result from more than one genotype.

Although genes do not determine complex phenotypes owing to interacting genes and environmental factors, genes do determine molecular phenotypes. Most genes specify the composition of a single **polypeptide chain.** All proteins are composed of one or more polypeptide chains aggregated together. Hemoglobin, the oxygen carrying protein in red blood cells, consists of two copies of a polypeptide chain denoted alpha-globin aggregated with two copies of a different polypeptide chain denoted beta-globin. The composition of these polypeptide chains is determined by the alpha-globin gene and the beta-globin gene, respectively, so these genes determine the hemoglobin phenotype.

Genes can exist in different forms or states. These alternative forms of a gene are called **alleles.** Different alleles code for somewhat different polypeptide chains. For example, one mutant form of the beta-globin gene encodes an aberrant polypeptide chain that tends to form crystals under low oxygen tension. These cause the red blood cells to collapse into half-moon or sickle shapes, yielding the name of the associated blood disease **sickle-cell anemia.** The consequences of the chronic red-cell destruction and reduced oxygen-carrying capability are severe. Yet the disease is maintained in some populations at relatively high frequency because heterozygous carriers are resistant to malaria (Chapter 2).

Gene Expression

The essentials of gene expression in the cells of higher organisms (eukaryotes) are outlined in Figure 1.1. From a biochemical point of view, a gene corresponds to a specific sequence of constituents (called nucleotides) along a molecule of DNA (deoxyribonucleic acid). There are four nucleotides in DNA, abbreviated according to the identity of the nitrogenous **base** that each contains: A (adenine), G (guanine), T (thymine), or C (cytosine). DNA is the genetic material. Different sequences of nucleotides that may occur in a gene,

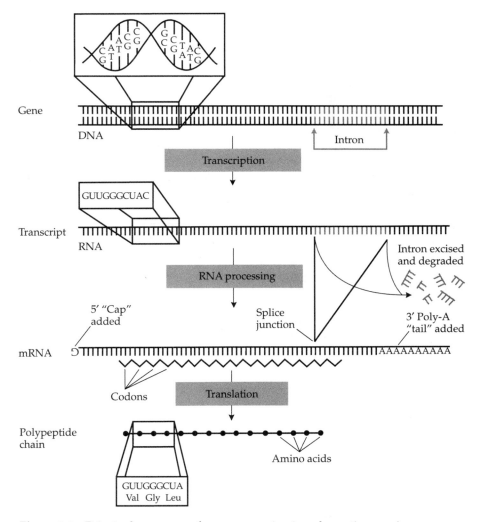

Figure 1.1 Principal processes of gene expression in eukaryotic organisms.

therefore, represent alleles. DNA molecules normally consist of two complementary helical strands held together by pairing between the bases: A in one strand is paired with T in the other strand across the way, and G in one strand is paired with C in the other.

The first step in gene expression is *transcription,* in which the sequence of nucleotides present in one DNA strand of a gene is faithfully copied into the nucleotides of an RNA molecule. As the RNA transcript is synthesized, each base in the DNA undergoes pairing with a base in an RNA nucleotide, which is then added to the growing RNA strand. The base-pairing rules are the same as those in DNA, except that in RNA nucleotides the base U (uracil) is found instead of T. The second step of gene expression is *RNA processing,* in which certain segments of the RNA transcript are removed by splicing. The segments that are eliminated are known as intervening sequences or **introns.** In RNA splicing, each intron is cleaved at its ends and discarded, while the ends of the flanking RNA sequences are joined together. The regions between the introns that remain in the fully processed RNA are known as **exons.** In addition to splicing of the exons, RNA processing also includes modifications to both ends of the RNA transcript. The fully processed RNA constitutes the **messenger RNA.** The messenger RNA undergoes *translation* on ribosomes in the cytoplasm to produce the polypeptide that is encoded in the sequence of nucleotides. In the translated part of the messenger RNA, each adjacent group of three nucleotides constitutes a coding group or **codon,** which specifies a corresponding amino acid subunit in the polypeptide chain. The standard **genetic code** showing which codons specify which amino acids is given in Table 1.1. The three-letter and one-letter abbreviations are both established conventions. The codon AUG (boxed in Table 1.1) specifies methionine and also serves as the start codon for polypeptide synthesis. Any of three codons—UAA, UAG, or UGA—specifies the end, or termination, of polypeptide synthesis, upon which the completed polypeptide chain is released from the ribosome.

The totality of DNA in a cell is the **genome.** Within a cell, genes are arranged in linear order along microscopic threadlike bodies called **chromosomes.** Each human reproductive cell contains one complete set of 23 chromosomes and has a genome size of approximately 3×10^9 base pairs. A typical chromosome contains several thousand genes, with humans averaging approximately 3500 genes per chromosome. The position of a gene along a chromosome is called the **locus** of the gene. In most higher organisms, as in human beings, each individual cell contains two copies of each type of chromosome, one inherited from its mother through the egg and one inherited from its father through the sperm. At any locus, therefore, every individual contains two alleles—one at each corresponding (homologous) position in the maternal and paternal chromosome. If the two alleles at a locus are indistinguishable in their effects on the organism, then the individual is said to be **homozygous** at the locus under consideration. If the two alleles at a locus are

Table 1.1 The standard genetic code

First nucleotide in codon (5' end)	Second nucleotide in codon				Third nucleotide in codon (3' end)
	U	C	A	G	
U	UUU Phe/F	UCU Ser/S	UAU Tyr/Y	UGU Cys/C	U
	UUC Phe/F	UCC Ser/S	UAC Tyr/Y	UGC Cys/C	C
	UUA Leu/L	UCA Ser/S	UAA Stop	UGA Stop	A
	UUG Leu/L	UCG Ser/S	UAG Stop	UGG Trp/W	G
C	CUU Leu/L	CCU Pro/P	CAU His/H	CGU Arg/R	U
	CUC Leu/L	CCC Pro/P	CAC His/H	CGC Arg/R	C
	CUA Leu/L	CCA Pro/P	CAA Gln/Q	CGA Arg/R	A
	CUG Leu/L	CCG Pro/P	CAG Gln/Q	CGG Arg/R	G
A	AUU Ile/I	ACU Thr/T	AAU Asn/N	AGU Ser/S	U
	AUC Ile/I	ACC Thr/T	AAC Asn/N	AGC Ser/S	C
	AUA Ile/I	ACA Thr/T	AAA Lys/K	AGA Arg/R	A
	AUG Met/M	ACG Thr/T	AAG Lys/K	AGG Arg/R	G
G	GUU Val/V	GCU Ala/A	GAU Asp/D	GGU Gly/G	U
	GUC Val/V	GCC Ala/A	GAC Asp/D	GGC Gly/G	C
	GUA Val/V	GCA Ala/A	GAA Glu/E	GGA Gly/G	A
	GUG Val/V	GCG Ala/A	GAG Glu/E	GGG Gly/G	G

distinguishable because of their differing effects on the organism, then the individual is said to be **heterozygous** at the locus. Typographically, genes are indicated in italics, and alleles are typically distinguished by uppercase or lowercase letters (*A* versus *a*), subscripts (A_1 versus A_2), superscripts (a^+ versus a^-), or sometimes just + and −. Using these symbols, homozygous individuals would be portrayed by any of these formulas: *AA*, *aa*, A_1A_1, A_2A_2, a^+a^+, a^-a^-, +/+, or −/−. As in the last two examples, the slash is sometimes used to separate alleles present in homologous chromosomes to avoid ambiguity. Heterozygous individuals would be portrayed by any of the formulas *Aa*, A_1A_2, a^+a^-, or +/−. The essence of Mendelian genetics is the *principle of segregation*, which states that each reproductive cell from a heterozygous individual contains only one of the two alleles, and that overall the reproductive cells contain the two alleles at equal frequency.

DNA Cleavage, Manipulation, and Amplification

Techniques for the experimental manipulation of DNA fragments are widely used for the study of genetic variation in natural populations. They take advantage of the unique features of DNA structure and replication. Some familiarity with these methods is necessary to understand how they are used to identify different types of genetic polymorphisms.

DNA hybridization. We have already noted that DNA normally consists of two intertwined helical strands in which the bases are paired, A with T and G with C. The individual strands can be separated (denatured) by heat or chemical treatment, and when returned to normal conditions the strands come back together again (renaturation) owing to the A–T and G–C attractions between the bases (Figure 1.2). The strands need not renature with their original pairing partner. Each strand can undergo hybridization with any sequence of complementary bases, provided that the stretch of complementary bases is long enough and perfectly matched enough to form a stable duplex. In particular, when DNA strands are denatured in the presence of an excess quantity of a radioactively labeled DNA fragment that is complementary in base sequence to part of the original duplex, the denatured strands will mainly hybridize with the radioactive DNA and thereby become radioactive themselves. Hybridization with such **probe DNA,** labeled with radioactivity or other chemical means, is the primary method of locating a particular unlabeled DNA fragment that has been immobilized on some solid supporting material, such as a disk of nitrocellulose filter paper.

DNA cleavage. Most methods for isolating double-stranded DNA shear the molecules into fragments averaging about 50 kb in length. (The abbreviation kb stands for **kilobase,** a unit of length equal to 1000 base pairs of double-stranded DNA or 1000 bases of single-stranded DNA.) The shearing takes place at random, so that any particular DNA sequence of interest (the

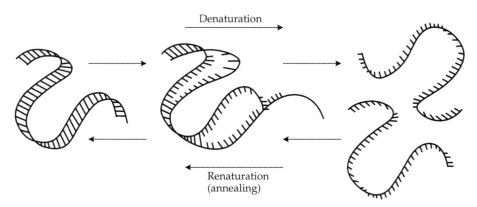

Denaturation

Renaturation
(annealing)

Figure 1.2 Denaturation and renaturation of the base pairs in a DNA duplex.

target sequence) will be present in a large set of random fragments, each differing in size and in the position of the target sequence within it. The problem of random shearing can be overcome by the use of special enzymes, called **restriction enzymes,** to digest each random fragment containing the target sequence at exactly the same positions relative to the target sequence. In this manner, a set of DNA fragments is produced that are of exactly the same size and contain the target sequence at exactly the same position.

A restriction enzyme cleaves double-stranded DNA at all sites in the molecule where the base sequence matches a particular short nucleotide sequence called the **restriction site** of the enzyme. Restriction enzymes occur naturally in bacteria. More than a hundred different types of restriction enzymes, each with its own restriction site, are commercially available. Some examples of restriction enzymes and their restriction sites are shown in Figure 1.3. In each strand, cleavage occurs at the position of the arrowhead. For example, the enzyme *Alu*I cuts DNA at sites containing the four-base sequence AGCT, and each strand is cleaved between the G and the C. By contrast, *Eco*RI cuts at the six-base sequence GAATTC, and each strand is cleaved between the G and the A. In naming restriction enzymes, the first three letters are italicized because they abbreviate the organism in which the enzyme was first discovered, for example, the *Eco* in *Eco*RI stands for *Escherichia coli.*

Figure 1.3 uses the symbols 5′ and 3′ to designate opposite ends of each DNA strand. These symbols reflect the ordered geometry of the strand. In essence, if a nucleotide were represented as a right arrow, →, then the geometry of DNA would require that the base available for pairing be attached to the shank of the arrow from below; likewise, if a nucleotide were represented as a left arrow, ←, then the geometry requires that the base available for pairing be attached to the shank of the arrow from above. In order for the bases in two DNA strands to be able to form pairs, one strand has to have all its bases jutting off toward the bottom, and the other strand has to have all its bases jutting off toward the top. This means that a strand with polarity →→→→→→ must be paired with a strand with polarity ←←←←←←,

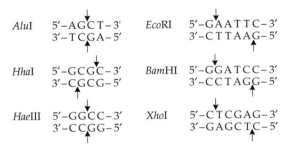

Figure 1.3 Restriction sites of several restriction enzymes.

because the first strand has all its bases pointing down, whereas the second has all its bases pointing up. In other words, the paired strands in a DNA duplex must point in opposite directions. The ends of each nucleotide strand have distinct chemical identities indicated by the symbols 5′ and 3′. For a DNA strand indicated as →→→→→→, the left-hand end is called the **5′ end,** and the right-hand end is called the **3′ end.** Likewise, for a DNA strand indicated as ←←←←←←, the left-hand end is the 3′ end and the right-hand end is the 5′ end. (In representing DNA, the convention is to place the 5′ end to the left for a single-stranded sequence, or at the upper left for a double-stranded sequence.) In the restriction sites in Figure 1.3, the top strand has the →→→→ orientation and the bottom strand the ←←←← orientation. As is the case with the examples in Figure 1.3, most restriction sites consist of a base sequence that is identical in the complementary strands, when the opposite polarity is taken into account. For example, 5′-AGCT-3′ is identical to 3′-TCGA-5′, except it is written in the reverse order.

Electrophoresis.   Electrophoresis is the separation of charged molecules in an electric field. DNA fragments differing in size move through an electric field at different rates. Each DNA sample is placed in a small slot near the edge of a rectangular slab of a jellylike material (usually agarose, a derivative of agar) about the size of a news magazine. The gel is placed inside a plastic box, each end of which is a chamber containing an electrode and a buffered solution (Figure 1.4) and an electric current is applied across the gel for several hours. DNA fragments in the samples move through the gel in response to the electric field, each fragment migrating a total distance proportional to the logarithm of its size. If a sample contains a limited number

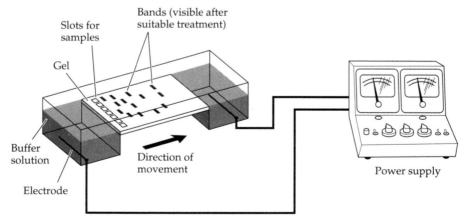

Figure 1.4   Apparatus for electrophoresis.

of DNA fragment sizes, the position of each fragment size can be located in the gel by viewing the gel under ultraviolet light, where the concentration of each fragment size produces a bright fluorescent **band** across the gel.

The Southern blot. DNA fragments that contain any particular sequence of bases can be identified by DNA hybridization using a probe that has a complementary base sequence. In the Southern blot procedure (named after its inventor), DNA restriction fragments that have been separated by electrophoresis are transferred (blotted) onto a sheet of nitrocellulose or nylon filter paper (Figure 1.5). The filter is heated in order to denature the two complementary DNA strands making up each molecule, and the strands are attached to the filter and immobilized by treatment with chemicals or ultraviolet light. Then the filter is bathed in a solution containing probe DNA that has been rendered radioactive. As the solution cools, the probe DNA strands hybridize with their complementary counterparts on the filter to form double-stranded molecules, and careful washing removes all of the probe DNA that has remained unhybridized. The filter is overlaid with a sheet of photographic film, where radioactive disintegrations from the bound probe result in visible bands.

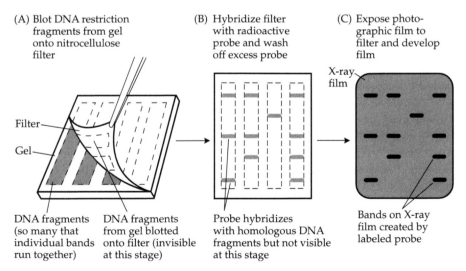

(A) Blot DNA restriction fragments from gel onto nitrocellulose filter

Filter

Gel

DNA fragments (so many that individual bands run together)　DNA fragments from gel blotted onto filter (invisible at this stage)

(B) Hybridize filter with radioactive probe and wash off excess probe

Probe hybridizes with homologous DNA fragments but not visible at this stage

(C) Expose photographic film to filter and develop film

X-ray film

Bands on X-ray film created by labeled probe

Figure 1.5 Southern blot procedure. (A) DNA fragments separated by electrophoresis are transferred and chemically attached to a filter. (B) The filter is mixed with radioactive probe DNA, which hybridizes with homologous DNA molecules on the filter. (C) After washing, the filter is exposed to photographic film, which develops dark bands caused by radioactive emissions from the probe.

DNA amplification. Particular DNA fragments can be amplified, or made so abundant that they can be visualized without the use of radioactivity or any other type of labeled probe. Amplification takes advantage of the fact that the DNA polymerase enzyme that catalyzes the synthesis of a new DNA strand cannot initiate strand synthesis on its own, but requires a short, single-stranded primer as a substrate to be elongated. The primer can consist of a short stretch of synthetic DNA (an **oligonucleotide**), typically about 20 bases in length. DNA polymerase also has the feature that a new DNA strand can be synthesized only by the addition of successive nucleotides to the 3' end of the primer. This means that any region of duplex DNA up to about 5 kb can be amplified using two oligonucleotide primers complementary in sequence to the termini of the region to be amplified (Figure 1.6). The template DNA strands are separated by heat, and renaturation is allowed to take place in the presence of an excess of the primer sequences. One primer initiates synthesis at the extreme right of the region, annealing to the 5' → 3' strand with its 3' end oriented toward the left; the other primer anneals at the extreme left of the complementary 3' ← 5' strand with its 3' end oriented toward the right. The polymerization reaction extends each primer at the 3' end, producing a new strand complementary to the template. One cycle of denaturation, primer annealing, and extension results in two replicas of each double-stranded template molecule that was present in the original mixture. In practice, the cycle is repeated about 25 times, each cycle doubling the number of replicas of the template. Because of the exponential growth in copy number, the amplification procedure is called the **polymerase chain reaction**, or **PCR.** The efficiency of PCR is greatly enhanced by the use of a

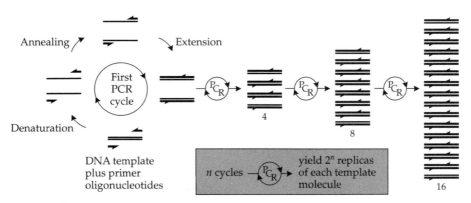

Figure 1.6 Polymerase chain reaction, in which repeated steps of DNA denaturation, primer annealing, and primer extension result in exponential growth in the number of copies of the DNA sequence flanked by the primers.

thermally stable DNA polymerase originally obtained from the thermophilic microorganism *Thermus aquaticus,* hence the name *Taq* **polymerase.** This polymerase allows PCR to be completely automated. The investigator mixes the reagents, inserts the reaction vessels into a programmable thermal cycler, and programs the machine to set the denaturation temperature, annealing temperature, extension time, number of cycles, and other parameters. To appreciate the power of PCR amplification, consider a target DNA sequence of 3 kb in a human reproductive cell containing 3×10^9 base pairs. Prior to amplification, the target sequence accounts for only 0.0001% of the total DNA. However, after 30 cycles of PCR amplification, the target sequence accounts for 99.9% of the total DNA and is sufficiently concentrated that, for most purposes, additional purification is unnecessary.

TYPES OF POLYMORPHISMS

One of the universal attributes of natural populations is phenotypic diversity. For most traits, many differing phenotypes can be found among the individuals in any population. Phenotypic diversity in many traits is impressive even with the most casual observation. Among human beings, for example, there is diversity with respect to height, weight, body conformation, hair color and texture, skin color, eye color, and many other physical and psychological attributes or skills. Population genetics must deal with this phenotypic diversity, and especially with that portion of the diversity which is caused by differences in genotype among individuals. Genetic variation, in the form of multiple alleles of many genes, exists in most natural populations. Next we consider some of the molecular methods by which genetic variation is revealed.

DNA Polymorphisms

The methods of DNA manipulation examined in the previous section can be used in a variety of combinations to analyze DNA from genomes sampled from natural populations. This is why anyone reading the literature in modern population genetics will encounter a bewildering variety of acronyms referring to different ways in which genetic variation is detected. Each approach has its own advantages and limitations. New approaches are continually being developed, and without some guidance the acronyms can be baffling. Currently, the most important types of variation at the DNA level are illustrated in Figure 1.7. These are examined next.

Figure 1.7A depicts a DNA duplex at high enough resolution to see the individual paired bases. One gene is represented along with its component parts. The *promoter* region contains sequences that control the time, tissue specificity, and level of transcription. The promoter is not transcribed but

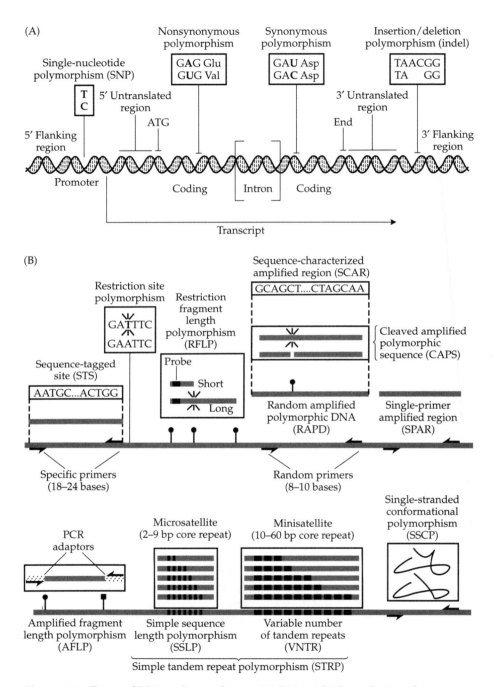

Figure 1.7 Types of DNA polymorphisms. (A) DNA at high resolution, showing polymorphisms at the level of individual base pairs. (B) DNA at lower resolution, showing polymorphisms at the level of restriction fragments or amplified fragments.

is part of the 5' flanking region of the gene. The transcript includes introns (removed during processing into the mRNA), a 5' untranslated region, a 3' untranslated region, and codons to start (AUG) and stop (UAA, UAG, or UGA) polypeptide synthesis. Several types of polymorphism are illustrated.

The acronym **SNP** means a **single-nucleotide polymorphism,** one in which individuals in the population may differ in the identity of the nucleotide pair present at a particular defined site in the DNA. In this example, the SNP is in the 5' flanking region of the gene and is a T/C polymorphism. That is, at this position, some DNA molecules in the population have a T–A base pair whereas other DNA molecules have a C–G base pair. The SNP therefore defines two "alleles," for which there could be three genotypes, namely, homozygous T–A, homozygous C–G, or heterozygous with T–A in one molecule and C–G at the corresponding site in the homologous molecule. The word "allele" is in quotes because the SNP need not be in a coding sequence, or even in a gene. Available data for the human genome suggest that any two randomly chosen sequences are likely to differ at one SNP site about every 1000–3000 base pairs (bp) in protein-coding DNA and about one SNP site every 500–1000 bp in noncoding DNA (Chakravarty 1999).

A **nonsynonymous polymorphism** is a single-nucleotide polymorphisms present in the coding region that alters a codon to result in an amino acid replacement in the polypeptide chain. In the example shown, the polymorphic nucleotide substitution in the RNA is shown in boldface (**A** versus **U**), resulting in a polymorphism for the codons GAG versus GUG. The GAG codon specifies Glu (glutamic acid), whereas the GUG codon specifies Val (valine). A nonsynonymous polymorphism therefore results in an **amino acid polymorphism.** In the human genome, a Glu/Val polymorphism at amino acid position 6 in the beta-globin gene is responsible for sickle-cell anemia.

A **synonymous polymorphism** is a single-nucleotide polymorphism present in the coding region that produces a synonymous codon and so does not result in an amino acid replacement in the polypeptide chain. The example given is a GA**U** codon versus a GA**C** codon, both of which code for Asp (aspartic acid). Synonymous polymorphisms are sometimes called **silent polymorphisms** because they do not change the amino acid. The existence of silent polymorphisms should not be taken to imply that organisms are indifferent to which of the synonymous codons are used for any particular amino acid. In many organisms, certain codons are preferred, especially in mRNAs that encode highly abundant proteins. It appears that the main reason for codon usage bias is translational accuracy, but translational speed may also play a role (Akashi 1993, 1995; Hartl et al. 1994; Eyre-Walker 1996).

An **indel** is an insertion/deletion polymorphism. In the example, the indel size is 2 base pairs and is located in the 3' flanking region of the gene.

In most cases of indel polymorphisms, it is unclear which sequence is the ancestral sequence; hence it is not known whether the mutation that created the polymorphism was an insertion or a deletion. Although many indels are fewer than 10 base pairs, some are much larger. Among the largest, typically 1–5 kb, are insertions of **transposable elements,** specialized nucleotide sequences of various kinds that are present in the genomes of virtually all organisms and that are able to replicate and move from one position to another (Chapter 3).

Figure 1.7B depicts DNA at lower resolution than part A, and it deals primarily with DNA polymorphisms that affect DNA fragments raging from 300 to 3000 base pairs in length. The type of region denoted **STS** is a **sequence-tagged site,** a region of known sequence that is present once per haploid genome found in a reproductive cell (Olson et al. 1989). One advantage of an STS marker is that the ends of the known sequence can be used to design primer oligonucleotides to specifically amplify the sequence. Using these primers, PCR enables one to detect the presence of a particular STS in a DNA fragment (for example, one cloned in a bacterial cell) or in a mixture of DNA fragments. The main use of STS markers is to identify cloned DNA fragments that contain particular STS markers known to be in or near a gene of interest, such as a gene implicated in an inherited disease.

A **restriction site polymorphism** is one in which some DNA molecules in the population contain a particular restriction site whereas others lack it. The example shown is that of an *Eco*RI site, 5′-GAATTC-3′, which is lacking in homologous DNA molecules carrying the A → T base substitution (denoted by the "sunburst") that eliminates the restriction site. The most easily identified type of restriction site polymorphism is one that results in a change in the size of a restriction fragment; it is known as an **RFLP**, or **restriction fragment length polymorphism** (Botstein et al. 1980). In the RFLP illustrated in Figure 1.7B, the restriction sites along a DNA molecule are indicated as solid dots above pointers to their positions in the molecule. The polymorphism eliminates the restriction site in the middle (indicated by the sunburst). In a Southern blot using a probe that hybridizes with the left-hand end of the restriction fragment, a DNA molecule that has all three restriction sites yields a single *short* fragment, because cleavage at the middle site eliminates the distal part of the fragment that does not hybridize with the probe. On the other hand, absence of the middle restriction site results in a single *long* fragment that includes the distal part. The differing sizes of the restriction fragments defines two alleles, *short* and *long,* and so for this RFLP there could be three possible genotypes, namely, *short/short* and *long/long* homozygous genotypes and the *short/long* heterozygous genotype.

At one time RFLPs were the principal molecular technique for identifying genetic polymorphisms, but the approach has several limitations. Chief among them are the need for sufficient genomic DNA from each of a large number of samples to do a Southern blot, the need for a probe (usually a short fragment of genomic DNA that has been cloned into a bacterial cell), and the need for radioactive label to achieve the most sensitive detection. For these reasons many population geneticists turned to methods based on PCR, so that DNA fragments from small amounts of genomic DNA could be amplified and detected.

One convenient method for identifying genetic polymorphisms is called **RAPD, or random amplified polymorphic DNA** (Welsh and McClelland 1990; Williams et al. 1990). It requires no probe DNA and no advance information about the genome of the organism, but uses a set of PCR primers of 8 to 10 bases whose sequence is random. The random primers are tried singly or in pairs in PCR reactions, and since the primers are so short, they often anneal to the template DNA at multiple sites. Some primers anneal in the proper orientation and at a suitable distance from each other to support amplification of the unknown sequence between them. Among the set of fragments are ones that can be amplified from some genomic DNA samples but not from others, which means that the presence or absence of the fragment is polymorphic in the population of organisms. In most organisms it is usually straightforward to identify a large number of RAPDs that can serve as genetic markers for many different kinds of genetic and population studies. An important feature of RAPDs and other detection methods based on PCR amplification is that presence of the fragment is dominant to absence of the fragment. In other words, if one allele (+) supports amplification but the alternative allele (−) does not, then DNA from the genotypes +/+ and +/− will support amplification equally well, whereas DNA from the genotype −/− will not support amplification. The + allele is therefore dominant to the − allele in regard to the corresponding RAPD fragment.

A whole panoply of different types of polymorphisms have come from the use of RAPD technology. If a fragment can be amplified using a single primer oligonucleotide, then it is called a **SPAR (single-primer amplified region).** When a particular amplified fragment is isolated and its nucleotide sequence determined, it becomes a **SCAR (sequence-characterized amplified region)** that can be converted into a conventional sequence-tagged site by the use of primers specific to the ends of the sequence. A particular amplified fragment may also include a restriction site polymorphism, in which case the polymorphism is called a **CAPS (cleaved amplified polymorphic site).** A CAPS is the analog of an RFLP, except that the genotype is identified by amplification and enzyme digestion, rather than in a Southern blot with a radioactive probe. This is why a CAPS is sometimes called a *PCR-RFLP.* Note that CAPS alleles, like RFLP alleles, are codominant, which means that the heterozygous genotype can be distinguished from each of the homozygous genotypes. For example, in Figure 1.7B, the amplified CAPS fragment is either cleaved into two pieces by the restriction enzyme or, if it has the mutant restriction site indicated by the sunburst, it is not cleaved. Hence, indicating the allele with the restriction site as + and that without the restriction site as –, then the genotype +/+ will yield two bands on amplification and cleavage, the genotype –/– will yield one band, and the genotype +/– will yield three bands. These differences enable all three genotypes to be distinguished.

Another class of amplified DNA polymorphisms is detected not with random primers, but rather with specific primers that are homologous to short double-stranded DNA sequences that have been attached to the ends of genomic restriction fragments using the enzyme DNA ligase. The attached sequences are known as **adaptors,** and they are indicated in Figure 1.7B as hatched bars at the ends of a genomic restriction fragment produced by cleavage with two different restriction enzymes (indicated by the filled circle and square). Cleavage of genomic DNA and adaptor ligation yields many different fragments flanked by adaptors, and all of the fragments are amplified simultaneously with the specific primers. The method is typically carried out with primers that amplify 50–75 fragments. As with RAPD, some of the fragments are polymorphic in whether genomic DNA samples will support their amplification. Such a polymorphism is known as an **AFLP,** which stands for **amplified fragment length polymorphism** (Vos et al. 1995). As in RAPD analysis, the alleles supporting amplification of AFLP fragments are dominant, which means that a single + allele is sufficient to support amplification, and so homozygous +/+ and heterozygous +/– genotypes cannot be distinguished.

The microsatellite and minisatellite polymorphisms in Figure 1.7 are based on short sequences that are repeated in tandem at one or more places in the genome. A **microsatellite** polymorphism has a very short core repeating unit of 2 to 9 base pairs. An example is the repeat

$$5'-CATCATCATCATCAT \cdot \cdot \cdot CATCATCATCATCAT-3'$$
$$3'-GTAGTAGTAGTAGTA \cdot \cdot \cdot GTAGTAGTAGTAGTA-5'$$

which can also be denoted as $(5'-CAT-3')_n$, where n is the number of repeats in the microsatellite. A **minisatellite** has the same sort of repeating structure, but the core unit is longer, typically 10 to 60 base pairs. In particular, microsatellite repeats may be present at many different locations in the genome, each flanked by restriction sites whose distance from the core repeat differs from one location to the next. Hence, if genomic DNA is cleaved with a restriction enzyme and the resulting fragments are separated by electropho

resis and hybridized in a Southern blot with a probe consisting of core repeats, each location in the genome containing the core repeats yields a separate band in the gel. Probes for some core repeats may hybridize with 100 or
more bands, yielding a bar-code sort of pattern often referred to as a **DNA
fingerprint** (Jeffreys et al. 1985)

For both microsatellites and minisatellites, each location in a chromosome that contains core repeats may have a different number (n) of copies of
the repeat. This means that at any particular location, a microsatellite or min-
isatelllite repeat has **multiple alleles** in the population, where each allele differs in its value of n. However, the alleles of different loci cannot be distinguished by probing with the core repeat because each allele yields a band
that is uninformative as to its origin in the genome. On the other hand, the
fragments derived from each genomic location can be assayed separately
using a probe that hybridizes with the unique flanking sequence of the
repeat, or if PCR primers to the unique flanking sequences are used to
amplify the repeat. With this approach, each allele yields a different fragment size due to the differing number of core repeats, and therefore yields a
distinct band in the gel. Genotypes that are heterozygous yield two bands,
and those that are homozygous yield one band. The alleles are therefore
codominant. When identified by unique-sequence probes or primers, a
microsatellite polymorphism is called a **simple sequence length polymorphism (SSLP)** and a minisatellite polymorphism is called a **variable number
of tandem repeats (VNTR)**. Generically, a polymorphism based on differences in the number of tandem repeats at a genomic location is called a **simple tandem repeat polymorphism**, or **STRP**.

The final type of DNA polymorphism illustrated in Figure 1.7 is a **single-
stranded conformational polymorphism**, or **SSCP**. This type of polymorphism results from the tendency for single-stranded DNA to fold back upon
itself to form a complex, three-dimensional conformation. In short molecules on the order of 300 base pairs, even a single nucleotide difference may
change the conformation of the single strands enough to change the electrophoretic mobility. Based on this phenomenon, regions of about 300 base
pairs that are amplified from genomic DNA are denatured into their single
strands, and the single strands are subjected to electrophoresis under special
conditions that do not allow them to renature. An SSCP is a polymorphism
within the fragment amplified from different organisms in a population that
causes the single strands to differ in electrophoretic mobility. SSCP is therefore one method for detecting single-nucleotide polymorphisms in any region of the genome, without the need for sequencing the homologous DNA
fragment isolated from a large number of individuals. The alleles underlying SSCP polymorphisms are codominant, so heterozygous genotypes can
be identified.

Uses of DNA Polymorphisms

Why are population geneticists interested in genetic polymorphisms? The interest can be justified on any number of grounds, but the following are the reasons most often cited. Each rationale is sufficient in its own right, and most are so self-evidently important as to require no further elaboration.

- To estimate the level of genetic variation in diverse populations of organisms differing in genetic organization (prokaryotes, eukaryotes, organelles), population size, breeding structure, or life-history characters.
- To examine and understand the patterns in which different types of genetic variation (e.g., synonymous versus nonsynonymous polymorphisms) occur throughout the genome.
- To understand the evolutionary mechanisms by which genetic variation is maintained, and the processes by which genetic polymorphisms within species become transformed into genetic divergence between species.
- To use genetic differences as a means of DNA fingerprinting to be able to uniquely identify different individuals in a population for purposes of criminal investigation, personal identification, determination of genetic relatedness, tracking of viral and bacterial epidemic diseases, and so forth.
- To use genetic polymorphisms as genetic markers in pedigree studies to identify, by genetic linkage, genes that are risk factors for inherited diseases in human populations or that are associated with favorable traits in domesticated animals and cultivated plants.
- To monitor the level of genetic diversity present in key indicator species present in biological communities in habitats exposed to chemical, biological, or physical stress.
- To use genetic polymorphisms within subpopulations of a species as indicators of population history, patterns of migration, and so forth.
- To understand the evolutionary origin, global expansion, and diversification of the human population.
- To understand the wild ancestors and practices of artificial selection in the origin of breeds of domesticated animals and cultivated plants.
- To analyze genetic differences between species in order to determine the ancestral history (phylogeny) of the species, and to trace the origin of morphological, behavioral, and other types of adaptations.

Protein Polymorphisms

Protein molecules can also be separated by electrophoresis. In **enzyme electrophoresis,** the position to which any particular enzyme has migrated is identified by immersing the gel in a solution containing a substrate for the enzyme along with a dye that precipitates where the enzyme-catalyzed reaction takes place. In this way the position of an enzyme in the gel is marked by the appearance of a dark band. Enzyme electrophoresis identifies a subset of all nonsynonymous nucleotide substitutions, because protein molecules of the same size that differ in charge due to an amino acid replacement can be separated. Polymorphisms of this type are called **allozymes**. Because the detection of allozyme polymorphisms requires a difference in amino acid sequence, there are fewer protein polymorphisms than DNA polymorphisms. Protein polymorphisms are usually more difficult to interpret than DNA polymorphisms, because a change in amino acid sequence of a protein might very well be expected to have some effect on the survival or reproduction of the organism (Lewontin 1991). Some allozyme polymorphisms are maintained because the heterozygous genotype has the highest average reproductive fitness. One well-established case is that of sickle-cell anemia in tropical Africa, where the near-lethality of the sickle-cell homozygotes is offset by the malaria resistance of the heterozygotes. Another example is an allozyme polymorphism for alcohol dehydrogenase in *Drosophila melanogaster* (Kreitman and Hudson 1991), but in this case the physiological basis of the heterozygote superiority is unclear. On the other hand, heterozygote superiority cannot explain the high incidence of allozyme polymorphisms in certain bacterial populations, such as the intestinal bacterium *Escherichia coli* (Whittam et al. 1983), since these organisms are haploid and therefore heterozygous, genotypes do not exist. In this organism many of the amino acid polymorphisms appear to be slightly detrimental (Sawyer et al. 1987).

Polymorphism for allozymes is demonstrated in Figure 1.8, which summarizes the results of electrophoretic surveys of from 14 to 71 enzymes (mostly around 20) in populations of 243 species. The numbers in parentheses are the number of species examined in each type of organism. The vertical axis, $\langle P \rangle$, refers to the estimated proportion of genes that are polymorphic; and the horizontal axis, $\langle H \rangle$, refers to the estimated proportion of enzyme-coding genes expected to be heterozygous in an average individual.

The vertical and horizontal bars indicate the limits within which 68% of *Drosophila* species are expected to fall. Thus, 68% of *Drosophila* species are expected have a proportion of polymorphic enzymes in the range 0.30–0.56 and an average heterozygosity for enzyme genes in the range 0.09–0.l9. Such bars could be attached to each point, but their lengths would be comparable to those for *Drosophila*. Because of the large variability in polymorphism and heterozygosity found within each group of organisms, Figure 1.8 admits of no simple summary. On the whole, there is a positive relationship between amount of polymorphism and degree of heterozygosity, which is as expected because the greater the fraction of polymorphic genes in a population, the more genes that are expected to be heterozygous in an average individual. The overall mean (average) of $\langle P \rangle$ in Figure 1.8 is 0.26 ± 0.15, and the mean of $\langle H \rangle$ is 0.07 ± 0.05. Vertebrates have the lowest average amount of genetic variation among the groups in Figure 1.8, plants come next, and inverte-

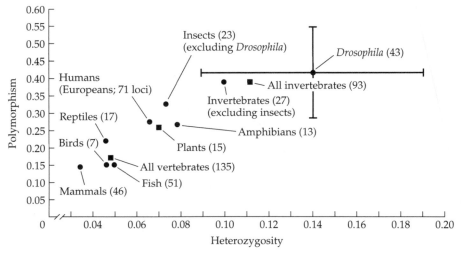

Figure 1.8 Estimated levels of heterozygosity and proportion of polymorphic loci derived from allozyme studies of various groups of plants and animals. The numbers are the number of species studied. Squares denote averages for plants, invertebrates, and vertebrates. Error bars are shown for *Drosophila* only. (Data from Nevo 1978.)

brates have the highest. *Drosophila* is the most genetically variable group of higher organisms so far studied, and mammals the least variable. Humans are fairly typical of large mammals. Extensive electrophoretic surveys among human populations yield $P \approx 0.30$ and $H \approx 0.05$.

Although allozyme polymorphisms are widespread, they are not universal. For example, both major subspecies of the cheetah (*Acinonynx jubatus*) are virtually monomorphic. For the East African subspecies (*A. j. raineyi*), estimates of P and H are 0.04 and 0.01, respectively; while for the South African subspecies (*A. j. jubatus*), the estimates are 0.02 and 0.0004, respectively. Most unusual is the finding of skin-graft acceptance between unrelated cheetahs from the South African subspecies. Graft acceptance means that the cheetah population is essentially homozygous for the major histocompatibility locus, which is abundantly polymorphic in other mammals. The cheetah, which was worldwide in its range at one time but presently numbers less than 20,000 animals, evidently underwent at least one very severe constriction in population number at some time in the geologically recent past, probably no later than 10,000 to 12,000 years ago (O'Brien et al. 1987).

Gene and Genotype Frequencies

At this point it is necessary to be a little more precise about the meaning of the term *polymorphism* as used with reference to Figure 1.8. A **polymorphic**

gene is typically defined as a gene for which there is a greater than 99% probability of observing more than one allele in a sample of 100 genes (50 diploid organisms). Conversely, a **monomorphic gene** is one that is not polymorphic. In any large population, rare alleles occur in virtually every gene. In human populations, the frequency of genotypes that are heterozygous for rare allozyme alleles is 1–2 per thousand individuals (Harris and Hopkinson 1972). These genes are regarded as monomorphic.

Alleles in natural populations usually differ in frequency from one allele to the next. The **allele frequency** of a prescribed allele among a group of individuals is defined as the proportion of all alleles at the locus that are of the prescribed type. The frequency of any prescribed allele in a sample is therefore equal to twice the number of genotypes homozygous for the allele (because each homozygote carries two copies of the allele), plus the number of genotypes heterozygous for the allele (because each heterozygote carries one copy), divided by two times the total number of individuals in the sample (because each individual carries two alleles at the locus).

For a codominant polymorphism, such as an indel, an RFLP, or an allozyme polymorphism, the allele frequencies can be estimated directly because each genotype can be separately identified. Consider, for example, a 32-base-pair indel found in the human chemokine receptor gene *CCR5*. This gene encodes a major macrophage coreceptor for the human immunodeficiency virus HIV-1, which is the causative agent of AIDS. Genotypes that are homozygous for the *CCDR5-Δ32* deletion are strongly resistant to infection by HIV-1. The *Δ32* allele is found in virtually all European populations, but the allele frequency varies (Lucotte and Mercier 1998). In one sample of 294 Parisians studied for the + (nondeletion) and *Δ32* (deletion) alleles, the numbers of individuals with each genotype were as follows:

+/+: 224 people   +/*Δ32*: 64 people   *Δ32*/*Δ32*: 6 people

Expressed as proportions these become the **genotype frequencies:**

+/+: 224/294 = 0.762   +/*Δ32*: 64/294 = 0.218   *Δ32*/*Δ32*: 6/294 = 0.020

Therefore, as estimated from this sample, the allele frequency of the + allele is:

$$\langle \text{Frequency of + allele} \rangle = (2 \times 224 + 64)/(2 \times 294) = 0.871$$

whereas the allele frequency of the *Δ32* allele is

$$\langle \text{Frequency of } \textit{Δ32} \text{ allele} \rangle = (2 \times 6 + 64)/(2 \times 294) = 0.129$$

Because they are proportions, the sum of the genotype frequencies, as well as that of the allele frequencies, must equal 1. The angled brackets are present to emphasize that the allele frequency in a sample of individuals from a population is only an estimate of the true allele frequency in the whole population, but the estimate will usually be close to the true frequency if the sample is sufficiently large. It is for this reason that allele frequency estimates should be based on samples of 100 or more individuals whenever possible. Since the term *gene* is sometimes used as a synonym for *allele*, the expression **gene frequency** is sometimes used as a synonym for *allele frequency* if the particular allele is clear from the context.

More generally, suppose that among n individuals sampled from a population the numbers of *AA, Aa,* and *aa* genotypes are n_{AA}, n_{Aa}, and n_{aa}, respectively. Following convention, we let p and q represent the allele frequencies of A and a, respectively, with $p + q = 1$. The estimate $\langle p \rangle$ of the allele frequency p in the population that was sampled is

$$\langle p \rangle = (2n_{AA} + n_{Aa})/2n \tag{1.1}$$

and the estimated sampling variance is

$$\langle \mathrm{Var}\langle p \rangle \rangle = \langle p \rangle (1 - \langle p \rangle)/2n \tag{1.2}$$

Equations 1.1 and 1.2 make use of several important concepts in statistics. Quantities used in describing populations are **parameters.** Although the exact values of parameters are usually unknown, their values can be estimated using samples from the population. In this book, whenever it is necessary to distinguish parameters from their estimates, we use unembellished symbols for parameters, for example p for the (unknown) frequency of an allele in a specified population, and the same symbol in angular brackets for the estimated value, in this example $\langle p \rangle$. The variance of an estimate is used for judging the reliability of the estimate. Since the variance is also estimated from data in the sample, the estimated variance is $\langle \mathrm{Var}\langle p \rangle \rangle$. The square root of the variance of an estimate is known as the **standard error** of the estimate.

The estimate in Equation 1.2 is the sampling variance of a binomial distribution. The binomial distribution occurs in such familiar probability contexts as a series of independent flips of a coin or rolls of a die. If the sampling and estimation were repeated many times, then approximately 68% of the estimates would fall within plus or minus one standard error of true value of the parameter, approximately 95% would fall within two standard errors, and approximately 99.7% would fall within three standard errors. These intervals are known as the 68%, 95%, and 99.7% **confidence intervals.** To take a specific example, 100 independent flips of an unbiased coin would be expected to yield a frequency of heads of 0.50 with a standard error of $\sqrt{[(0.5)(0.5)/100]} = 0.05$. Hence, approximately 68% of such trials would be expected to yield an observed frequency of heads in the range 0.45–0.55, 95% to yield an observed frequency in the range 0.40–0.60, and 99.7% to yield an observed frequency in the range 0.35–0.65. To put the matter in somewhat different terms, in a large number of such trials, 32% of the observed frequencies would be expected to differ from the true value by more than one standard error, 5% by more than two, and only 0.3% by more than three. These approximations all assume that the repeated estimates conform to the familiar, bell-shaped normal distribution.

As a population genetics example, consider again the *CCDR5-Δ32* deletion. A survey of 111 French Basques from the Pyrenees gave an estimate $\langle q \rangle = 0.018$ with a standard error of 0.009. This estimate of q is 8.2 standard errors below the average across all European populations (Lucotte and Mercier 1998), and a deviation as great or greater occurring by chance alone has a probability of 1.7×10^{-16}. Clearly, the discrepancy cannot be attributed to sampling variation. In this case, the explanation is that the Basque population was founded around 18,000 years ago by relatively few migrants, who then diverged genetically from other European populations owing to their relative geographical isolation in the mountains. This scenario is supported by the analysis of allele frequencies at many loci (Bertranpetit et al. 1995) as well as analysis of mitochondrial DNA (Calafell and Bertranpetit 1994).

ORGANIZATION OF GENETIC VARIATION

The situation regarding the Basques illustrates an issue of general importance in the population genetics of virtually all organisms. Biological species almost always exhibit some sort of **geographical structure,** which means a nonrandom pattern in the spatial distribution of organisms. Members of a species are rarely distributed homogeneously in space. There is almost always some sort of clumping or aggregation, some schooling, flocking, herding, or colony formation. Population subdivision is often caused by environmental patchiness, areas of favorable habitat intermixed with unfavorable areas. Such environmental patchiness is obvious in the case of terrestrial organisms on islands in an archipelago, but patchiness is a common feature of most habitats—freshwater lakes have shallow and deep areas, meadows have marshy and dry areas, forests have sunny and shady areas. Population subdivision can also be caused by social behavior, as when wolves form packs. Even the human population is clumped or aggregated— into towns and cities, away from deserts and mountains.

Populations

In population genetics, the word *population* does not usually refer to an entire species but rather to a group of individuals of the same species living within a sufficiently restricted geographical area so that any member can potentially mate with any other member (provided, of course, that they are of the opposite sex). The focus is on the local interbreeding units of possibly large, geographically structured populations, because it is within such local units that systematic changes in allele frequency occur that ultimately result in the evolution of adaptive characteristics. Such local interbreeding units are often called **local populations** or **subpopulations.** In this book we use the word *population* to mean *local population*—the actual, evolving unit of a species—unless a broader meaning is clear from the context. Local populations are also sometimes referred to as **Mendelian populations** or **demes.**

Models

Equation 1.1 specifies how allele frequencies can be estimated for codominant alleles, because each genotype can be identified individually. The situation is more difficult with genetic markers that are dominant, for example, RAPDs or AFLPs, because homozygous and heterozygous genotypes that support PCR amplification of the genetic marker are indistinguishable. In these cases, one cannot estimate the allele frequencies without making some assumptions about the mathematical relations between allele frequencies and genotype frequencies. Such mathematical relations are usually inferred from models that specify the types and frequencies of matings that take place in the population.

A **model** is an intentional simplification of a complex situation designed to eliminate extraneous detail in order to focus on the essentials. In population genetics, we must contend with factors such as population size, patterns of mating, geographical distribution of individuals, mutation, migration, and natural selection (differential survival or reproductive success). Although we wish ultimately to understand the combined effects of all these factors and more, the factors are so numerous and interact in such complex ways that they cannot usually be grasped all at once. Simpler situations are therefore devised in which a few identifiable factors are the most important and others can be neglected.

One kind of model frequently used in population genetics is the **mathematical model,** which is a set of hypotheses that specifies the mathematical relations between measured or measurable quantities (the parameters) that characterize a population. Mathematical models can be extremely useful. They express concisely the hypothesized quantitative relationships between parameters. They reveal which parameters are the most important ones in a system and thereby suggest critical experiments or observations. They serve as guides to the collection, organization, and interpretation of observed data. And they make quantitative predictions about the behavior of a system that can, within limits, be confirmed or shown to be false. The validity of a model must, of course, be tested by determining whether the hypotheses on which it is based and the predictions that grow out of it are consistent with observations.

Mathematical models are always simpler than the actual situation they are designed to elucidate. Many features of the actual system are intentionally left out of the model, because to include every aspect of the system would make the model too complex and unwieldy. Construction of a model always involves compromise between realism and manageability. A completely realistic model is likely to be too complex to handle mathematically, and a model that is mathematically simple may be so unrealistic as to be useless. Ideally, a model should include all essential features of the system and exclude all nonessential ones. How good or useful a model is often depends on how closely this ideal is approximated. In short, a model is a sort of metaphor or analogy. Like all analogies, it is valid only within certain limits and, when pushed beyond these limits, becomes misleading or even absurd.

One of the most important mathematical models in population genetics deals with organisms with a very simple life history called **nonoverlapping generations,** in which the individuals in each generation die before the members of the next generation are born. The model applies literally only to annual plants (Figure 1.9) and some short-lived invertebrates. In such organisms, all members of any generation are born at the same time, mature and reach sexual maturity synchronously, mate simultaneously, and die immediately after

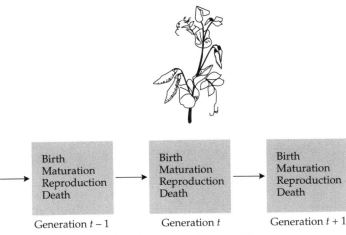

Figure 1.9 Nonoverlapping generation model.

producing the new generation. The key simplification is that, at any time, all members of the population are of the same age, and no individuals survive from one generation to the next. This model is often used in population genetics as a first approximation to populations that have more complex life histories. Although at first glance the model seems grossly oversimplified, calculations of expected genotype frequencies based on the model are adequate for many purposes, and they are often satisfactory first approximations even for populations with long and complex life histories such as in humans.

Random Mating

With nonoverlapping generations, the calculation of genotype frequencies from knowledge of allele frequencies is quite straightforward. The genotype frequencies are determined in part by the manner in which mating pairs are formed. Under **random mating,** mating takes place at random with respect to the genotypes under consideration, as if determined by random collisions. The probability of two genotypes forming a mating pair is therefore equal to the product of their respective genotype frequencies.

It is important to keep in mind that mating can be random with respect to some traits but nonrandom with respect to others in the same population. In human populations, for example, mating seems to be random with respect to most DNA polymorphisms, allozyme phenotypes, blood groups, and many other characteristics, but mating is nonrandom with respect to other traits such as skin color and height. Genotype frequencies are also influenced by various evolutionary forces including mutation, migration, and natural selection. For the moment, these evolutionary forces will be assumed to be absent or at least negligibly small in magnitude. Additionally, genotype frequencies are affected by chance statistical fluctuations that occur in all small populations, but for now we suppose that each local population is sufficiently large that small-population effects can be neglected.

The Hardy-Weinberg Principle

The main assumptions of the standard random-mating model are:

- Diploid organism
- Sexual reproduction
- Nonoverlapping generations
- Random mating
- Large population size
- Equal allele frequencies in the sexes
- No migration
- No mutation
- No selection

Under these assumptions, the genotype frequencies for a gene with two alleles can be deduced quite easily, as shown in Table 1.2. We assume that genotype frequencies of *AA*, *Aa*, and *aa* in the parental generation are *D*, *H*, and

Table 1.2 Demonstration of the Hardy-Weinberg principle

Mating	Frequency of mating	Offspring genotype frequencies		
		AA	*Aa*	*aa*
$AA \times AA$	D^2	1	0	0
$AA \times Aa$	$2DH$	$\frac{1}{2}$	$\frac{1}{2}$	0
$AA \times aa$	$2DR$	0	1	0
$Aa \times Aa$	H^2	$\frac{1}{4}$	$\frac{1}{2}$	$\frac{1}{4}$
$Aa \times aa$	$2HR$	0	$\frac{1}{2}$	$\frac{1}{2}$
$aa \times aa$	R^2	0	0	1
	Totals (next generation)	D'	H'	R'

where: $D' = D^2 + 2DH/2 + H^2/4 = (D + H/2)^2 = p^2$
$H' = 2DH/2 + 2DR + H^2/2 + 2HR/2 = 2(D + H/2)(R + H/2) = 2pq$
$R' = H^2/4 + 2HR/2 + R^2 = (R + H/2)^2 = q^2$

R, respectively, where $D + H + R = 1.0$. The allele frequencies of A and a are given by

$$p = (2D + H)/2 = D + H/2 \quad \text{and} \quad q = (2R + H)/2 = R + H/2 \quad (1.3)$$

These formulas follow immediately from Equation 1.1 when expressed in terms of genotype frequencies rather than absolute numbers. Note that $p + q = 1.0$, which is a consequence of the fact that the gene has only two alleles.

With three genotypes, there are six possible types of matings. When mating is random, these mating types occur in proportion to the genotypic frequencies in the population. For example, the mating $AA \times AA$ occurs only when an AA male mates with an AA female, and this occurs a proportion $D \times D$ (or D^2) of the time. Similarly, an $AA \times Aa$ mating occurs when an AA female mates with an Aa male (proportion $D \times H$), or when an Aa female mates with an AA male (proportion $H \times D$)—so the overall proportion of $AA \times Aa$ matings is $DH + HD = 2DH$. The frequencies of these and the other types of matings are given in the second column of Table 1.2.

The offspring genotypes produced by the matings are given in the last three columns of Table 1.2. The offspring frequencies follow from Mendel's law of segregation, which states that an Aa heterozygote produces an equal number of A-bearing and a-bearing gametes (**gamete** is a useful word meaning sperm or egg). Homozygous AA genotypes produce only A-bearing gametes and homozygous aa genotypes produce only a-bearing gametes. Therefore, a mating of AA with aa produces all Aa offspring, a mating of AA

with *Aa* produces 1/2 *AA* and 1/2 *Aa* offspring, a mating of *Aa* with *Aa* produces 1/4 *AA*, 1/2 *Aa*, and 1/4 *aa* offspring, and so forth.

The genotype frequencies of *AA*, *Aa*, and *aa* after one generation of random mating are denoted in Table 1.2 as D', H', and R' respectively. (Population geneticists commonly use unprimed and primed symbols to denote population parameters in two successive generations.) The new genotype frequencies are calculated as the sum of the cross products shown at the bottom of the table. The new genotype frequencies P', Q', and R' simplify to:

$$AA: D' = p^2 \qquad Aa: H' = 2pq \qquad aa: R' = q^2 \qquad (1.4)$$

a result known as the **Hardy-Weinberg principle** after Godfrey Hardy (1908) and Wilhelm Weinberg (1908).

The calculation in Table 1.2 illustrates the important principle that *random mating of individuals is usually equivalent to random union of gametes.* This equivalence means that we can cross-multiply the allele frequencies along the margins in the sort of square shown in Figure 1.10. As shown at the right of the figure, the cross multiplication easily leads to the Hardy-Weinberg frequencies. The square works because it is a systematic way of going through all the possibilities of gamete combination. The probability that a sperm or egg carries *A* is p; the probability that a sperm or egg carries *a* is q. With random combination of gametes, the chance that an *A*-bearing sperm fertilizes an *A*-bearing egg is $p \times p = p^2$, which is the frequency of *AA* genotypes. The probability that an *A*-bearing sperm fertilizes an *a*-bearing egg is $p \times q = pq$; and the probability that an *a*-bearing sperm fertilizes an *A*-bearing egg is $q \times p = qp$. Altogether the frequency of *Aa* heterozygotes is $pq + qp = 2pq$. Finally, the probability that an *a*-bearing sperm fertilizes an *a*-bearing egg is $q \times q = q^2$, which is the genotype frequency of *aa*. Note that $p^2 + 2pq + q^2 = (p + q)^2 = (1)^2 = 1$, thus accounting for all of the offspring.

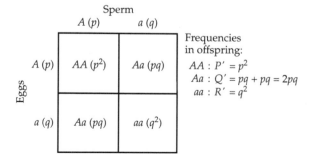

Figure 1.10 Cross-multiplication square showing Hardy-Weinberg frequencies resulting from random mating with two alleles.

Implications of Random Mating

The Hardy-Weinberg principle provides the foundation for many theoretical investigations in population genetics. One of the most important implications emerges when we calculate the allele frequencies p' and q' of A and a in the next generation. Using Equation 1.4, the allele frequencies of A and a are

$$p' = (2D' + H')/2 = (2p^2 + 2pq)/2 = p(p + q) = p$$
$$q' = (2R' + H')/2 = (2q^2 + 2pq)/2 = q(q + p) = q$$

In other words, the allele frequencies in the next generation are exactly the same as they were the generation before. This means that Mendelian inheritance, by itself, tends to keep the allele frequencies constant and to preserve genetic variation. Because the allele frequencies remain the same generation after generation, so do the genotype frequencies in the proportions p^2, $2pq$, and q^2, which is often called the **Hardy-Weinberg equilibrium (HWE).**

Chi-square Test for HWE

The mere fact that observed genotype frequencies may happen to fit HWE cannot be taken as evidence that all of the assumptions in the model are valid. The principle is not very sensitive to certain kinds of departures from the assumptions, particularly those pertaining to a very large population size with no migration, mutation, or selection. On the other hand, the relative insensitivity to departures from its assumptions gives the principle some robustness, because it implies that HWE can be valid to a first approximation even when one or more of the assumptions is violated.

The usual test for goodness of fit of observed data to HWE is a chi-square test. The test statistic is usually symbolized X^2, and under the hypothesis of HWE the X^2 has approximately a chi-square distribution. Application of the test can be illustrated using the sample of Parisians assayed for the CCDR5-$\Delta 32$ deletion polymorphism. There were 224 +/+ homozygotes, 64 +/$\Delta 32$ heterozygotes, and 6 $\Delta 32/\Delta 32$ homozygotes. The allele frequencies p for + and q for $\Delta 32$ were estimated earlier as $\langle p \rangle = 0.871$ and $\langle q \rangle = 0.129$. With HWE for these allele frequencies, the expected genotype frequencies are $(0.871)^2 = 0.758$, $2(0.871)(0.129) = 0.225$, and $(0.129)^2 = 0.017$. Multiplying each of these by the sample size (294 persons) gives the expected numbers as 222.9, 66.2, and 4.9. This conversion is necessary because the chi-square test must be based on the observed numbers, not ratios or proportions. The comparison is thus between the observed (*obs*) and expected (*exp*) numbers:

obs	224	64	6	Total = 294
exp	222.9	66.2	4.9	Total = 294

(Calculating the totals for the observed and expected numbers is a useful crosscheck.) In comparisons of this type, the value of X^2 is calculated as

$$X^2 = \sum \frac{(obs - exp)^2}{exp} \tag{1.5}$$

where the summation sign Σ means summation over all classes of data, in this case all three genotypes. The resulting value of

$$X^2 = \frac{(224 - 222.9)^2}{222.9} + \frac{(64 - 66.2)^2}{66.2} + \frac{(6 - 4.9)^2}{4.9} = 0.32$$

is the test statistic.

Associated with any X^2 value is a second number called the **degrees of freedom** for that X^2. In general, the number of degrees of freedom associated with a X^2 equals the number of classes of data (in this case, 3) minus 1 (because the totals must be equal), minus the number of parameters estimated from the data (in this case, 1, because the parameter p was estimated from the data), so the number of degrees of freedom for our chi-square value is $3 - 1 - 1 = 1$. (Note: A degree of freedom is not deducted for estimating q because of the relation $q = 1 - p$; that is, once p has been estimated, the estimate of q is automatically fixed, so we deduct just the one degree of freedom corresponding to p.)

The actual assessment of goodness of fit is determined from Figure 1.11. To use the chart, find the value of X^2 along the horizontal axis; then move vertically from this value until the proper degrees-of-freedom line is intersected; then move horizontally from this point of intersection to the vertical axis and read the corresponding probability value. In the present case, with $X^2 = 0.32$ and one degree of freedom, the corresponding probability value is about $P = 0.63$. This probability has the following interpretation: It is the probability that chance alone could produce a deviation between the observed and expected values at least as great as the deviation actually obtained. Thus, if the probability is large, it means that chance alone could account for the deviation, and it strengthens our confidence in the validity of the model used to obtain the expectations—in this case HWE. On the other hand, if the probability associated with the X^2 is small, it means that chance alone is not likely to lead to a deviation as large as actually obtained, and it undermines our confidence in the validity of the model. Where exactly the cutoff should be between a "large" probability and a "small" one is, of course, not obvious, but there is an established guideline to follow. If the probability is less than 0.05, then the result is said to be **statistically signifi-**

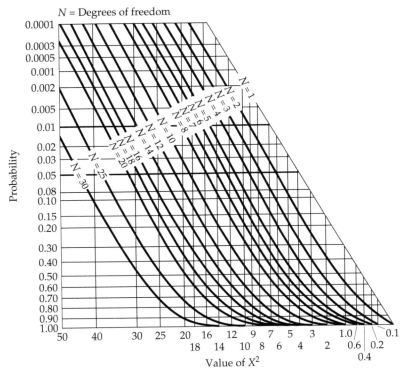

Figure 1.11 Graph for interpreting X^2 in tests for goodness of fit of observed to expected numbers. The probability value for any X^2 is the probability of obtaining a fit as bad or worse than that actually observed, under the assumption that the expected numbers are correct.

cant, and the goodness of fit is considered sufficiently poor that the model is judged invalid for the data. Alternatively, if the probability is greater than 0.05, the fit is considered sufficiently close that the model is not rejected. Because the probability in the $\Delta 32$ example is 0.63, which is considerably greater than 0.05, we have no reason to reject the hypothesis that the genotype frequencies are in HWE for this gene.

In contrast to the sample of people from Paris, that from Rheims does not fit HWE. In this sample the observed and expected values are

obs	234	36	6	Total = 276
exp	230.1	43.8	2.1	Total = 276

and the $X^2 = 8.8$ with an associated $P = 0.002$. A P value less than 0.01, as obtained in this case, is said to be **statistically highly significant,** and the hypothesis of HWE is decisively rejected. (Alternative goodness-of-fit tests are usually preferred when the smallest expected value is less than 5, but here the test is for illustrative purposes.) Why would this sample not fit HWE? In theory, any of the assumptions needed for HWE may be incorrect, but often a deficiency of heterozygotes, as observed in this case, results from combining samples from two or more subpopulations, all in HWE but differing in allele frequency. In this particular case, the observed numbers are consistent with an equal mixture of two HWE subpopulations having $\Delta 32$ allele frequencies of $q = 0.18$ and $q = 0$, which yields an X^2 of 2.1. There are many other possibilities also.

Recessive Alleles Hidden in Heterozygotes

HWE helps solve the dilemma that arises when studying polymorphisms of dominant genetic markers such as RAPD or AFLP polymorphisms. The dilemma is that homozygous and heterozygous genotypes cannot be distinguished, and so the allele frequencies cannot be estimated directly. The solution is that, if one is willing to assume HWE, then the allele frequencies can be estimated anyway. The trick is to use the observed frequency of *recessive homozygotes* to estimate the q^2 term of the HWE. That is, if R is the frequency of homozygous recessive genotypes found among a sample of n individuals, then $\langle q \rangle$ and its sampling variance are estimated as

$$\langle q \rangle = \sqrt{R} \qquad \langle \text{Var}\langle q \rangle \rangle = \frac{1-R}{4n} \tag{1.6}$$

To apply the formulas, consider a RAPD assayed in DNA from 100 individuals among which 75 support amplification and 25 do not. Then $R = 25/100$ and $\langle q \rangle = 0.50 \pm 0.043$. The "expected" genotype frequencies are therefore 25 $+/+$, 50 $+/-$, and 25 $-/-$. There is no chance for a goodness-of-fit test to HWE because there are 0 degrees of freedom, calculated as 2 (because there are two classes of data) $- 1$ (for the totals to add up) $- 1$ (for estimating the frequency of the recessive allele) $= 0$. The lack of any degrees of freedom is the reason why the expected frequencies fit the observed frequencies exactly.

One of the most important implications of HWE pertains to rare recessive alleles. When a recessive allele is rare, most individuals who carry the allele are heterozygous. This principle follows immediately from the genotype frequencies, because there are $2pq$ heterozygotes and q^2 recessive homozygotes. The ratio of heterozygous to homozygous recessive frequencies is $2pq/q^2 = 2p/q$. For example, if $q = 0.10$, the ratio $2p/q = 18$, which means that there are 18 times as many heterozygotes as recessive homozygotes in the population. A few other illustrative values for $(q, 2p/q)$ are (0.05, 38), (0.01, 198), (0.005, 398), (0.001, 1998). These values indicate that, as the recessive allele becomes increasingly rare, the greater becomes the proportion of heterozygous carriers relative to homozygous recessives.

To take a real example, consider the recessive disease cystic fibrosis, a severe condition associated with abnormal glandular secretions caused by mutations in the *CFTR* (cystic fibrosis transmembrane conductance regulator) gene in chromosome 7. Among Caucasians the incidence of affected individuals is approximately 1 in 2500 newborns, yielding the allele frequency estimate $\langle q \rangle = \sqrt{(1/2500)} = 0.02$. The frequency of heterozygotes is therefore estimated as $2(0.02)(1-0.02) = 0.0392$, or about 1 in 26 persons. This means that, although only 1 person in 2500 is actually affected with cystic fibrosis, about 1 person in 26 is a heterozygous carrier of a harmful mutation. Interestingly, approximately 70% of the mutant *CFTR* alleles have the so-called *ΔF508* deletion, a three-base-pair deletion of codon number 508 (Kerem et al. 1989).

Multiple Alleles and X-Linked Genes

Just as HWE for two alleles can be expressed formally as the binomial square

$$(p\ A + q\ a)^2 = p^2\ AA + 2pq\ Aa + q^2\ aa$$

HWE for multiple alleles can be expressed formally as the multinomial square

$$(p_1\ A_1 + p_2\ A_2 + \cdots + p_n\ A_n)^2 \tag{1.7}$$

where each p_i is the allele frequency of the allele A_i and n is the total number of alleles, hence $\Sigma p_i = 1$. Expanding Equation 1.7 into its component terms, we have

Genotype frequency of any homozygous genotype $A_iA_i = p_i^2$

Genotype frequency of any heterozygous genotype $A_iA_j = 2p_ip_j$ $\quad$ (1.8)

A systematic way to calculate the genotype frequencies with three alleles is shown in Figure 1.12. More alleles can be handled by increasing the number of squares.

Perhaps the most important practical application of HWE for multiple alleles is that in DNA typing for purposes of criminal identification. Typing agencies often assay SSLPs or VNTRs for this purpose; hence there are multiple alleles (National Research Council 1992, 1996). The potential power of the approach can be appreciated by considering a VNTR locus with 20 al-

Sperm

	$A_1\ (p_1)$	$A_2\ (p_2)$	$A_2\ (p_3)$	
$A_1\ (p_1)$	$A_1A_1\ (p_1{}^2)$	$A_1A_2\ (p_1p_2)$	$A_1A_3\ (p_1p_3)$	Frequencies in offspring: $A_1A_1 : p_1{}^2$ $A_1A_2 : p_1p_2 + p_1p_2 = 2p_1p_2$ $A_1A_3 : p_1p_3 + p_1p_3 = 2p_1p_3$ $A_2A_2 : p_1{}^2$ $A_2A_3 : p_2p_3 + p_2p_3 = 2p_2p_3$ $A_3A_3 : p_3{}^2$
$A_2\ (p_2)$	$A_1A_2\ (p_1p_2)$	$A_2A_2\ (p_2{}^2)$	$A_2A_3\ (p_2p_3)$	
$A_3\ (p_3)$	$A_1A_3\ (p_1p_3)$	$A_2A_3\ (p_2p_3)$	$A_3A_3\ (p_3{}^2)$	

(Eggs, left margin)

Figure 1.12   HWE for three alleles.

leles, each equally frequent (not usually the case, but the simplification is useful for illustration). With HWE, 95% of the persons in the population are heterozygous, and the frequency of any particular heterozygous genotype is $2(0.05)(0.05) = 1/200$. If a DNA sample from a crime scene matches that from a suspect at a number i of VNTR loci, each with 20 equally frequent alleles and independent of one another, then the probability of a perfect match at all i loci is given by the successive powers $(1/200)^i$. For $i = 1$ through 8, the powers are as follows, where the probability of an exact match of i loci is given after the implication sign ($\Rightarrow$):

$1 \Rightarrow 5.00 \times 10^{-3}$ $2 \Rightarrow 2.50 \times 10^{-5}$ $3 \Rightarrow 1.25 \times 10^{-7}$ $4 \Rightarrow 6.25 \times 10^{-10}$

$5 \Rightarrow 3.13 \times 10^{-12}$ $6 \Rightarrow 1.56 \times 10^{-14}$ $7 \Rightarrow 7.81 \times 10^{-17}$ $8 \Rightarrow 3.91 \times 10^{-19}$

Even though this is an artificial example, it serves to demonstrate that a simultaneous match at 6–8 highly polymorphic loci with multiple alleles makes a very strong case for identity between the suspect and the source of the criminal evidence, especially if the assumption of independence between the loci can be justified.

Another important special case of HWE concerns gene located in the X chromosome. In mammals and many insects, females have two copies of a chromosome designated "X," whereas males have one copy of the X chromosome and one copy of a different chromosome designated "Y." The X and Y chromosomes segregate (separate) from each other during the formation of sperm, so half the sperm from a male carry the X chromosome and half

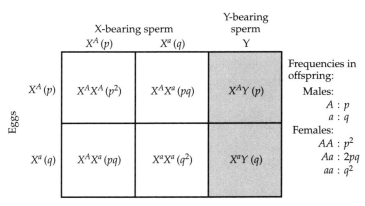

Figure 1.13   HWE for X-linked genes.

carry the Y chromosome. (In birds, moths, and butterflies, the sex-chromosome situation is the reverse: males are XX and females XY.) Although the Y chromosome carries very few genes other than those involved in the determination of sex and male fertility, the X chromosome carries as full a complement of genes as any other chromosome. Genes in the X chromosome are called **X-linked genes.** Because males have only one X chromosome, and because the Y chromosome lacks a homolog of most genes in the X, a recessive X-linked allele present in a male will be expressed phenotypically. For X-linked genes with two alleles, therefore, there are three female genotypes (*AA, Aa,* and *aa*) but only two male genotypes (*A* and *a*).

One of the important features of random mating for X-linked genes is that conditions due to a rare recessive allele will be more common in males than in females. The reason can be seen in Figure 1.13, which shows the consequences of random mating with two X-linked alleles when the allele frequencies are equal in the sexes. The alleles are denoted X^A and X^a. Note that in females the genotype frequencies equal the HWE, whereas in males the genotype frequencies equal the allele frequencies. The sex difference occurs because q, which equals the proportion of males with the recessive phenotype, will always be greater than q^2, which equals the proportion of females with the recessive phenotype. For example, with the X-linked "green" type of color blindness, $q \approx 0.05$, so the ratio of affected males to affected females is $q/q^2 = 1/q \approx 1/0.05 = 20$. For the X-linked "red" type of color blindness due to a closely linked gene, $q \approx 0.01$, so in this case the ratio of affected males to affected females is approximately $1/0.01 = 100$. The color-blindness examples show that as a recessive X-linked allele becomes more rare, the excess of affected males over affected females grows larger.

Multiple Loci: Linkage and Linkage Disequilibrium

Statistically, HWE means that the alleles present at a locus are in random association with each other in the genotypes. It therefore may seem paradoxical that two genes, A and B, present in the same population may each obey HWE individually, yet the alleles of A and B can remain in *nonrandom* association in the *gametes* that form each generation. That this is possible is shown in Figure 1.14. The gametic types are arrayed across the top for two alleles at each of two loci, A and B. The gametic frequencies are given in two completely equivalent ways. One is in terms of the parameters P_{11}, P_{12}, P_{21}, and P_{22} according to whether the alleles present in the gametes are $A_1 B_1$, $A_1 B_2$, $A_2 B_1$, or $A_2 B_2$. The other is in terms of the allele frequencies p_1 and p_2 of alleles A_1 and A_2, allele frequencies q_1 and q_2 of alleles B_1 and B_2, and a parameter designated D that measures the nonrandom association, called **linkage disequilibrium** or **LD**, between the loci. When $D = 0$ the gametic frequencies equal the products of the relevant allele frequencies, and the loci are said to be in **linkage equilibrium** or **LE**. The two types of parameterizations in Figure 1.14 are equivalent because of the definitions:

$$p_1 = P_{11} + P_{12} \qquad p_2 = P_{21} + P_{22}$$

$$q_1 = P_{11} + P_{21} \qquad q_2 = P_{12} + P_{22} \tag{1.9}$$

$$D = P_{11}P_{22} - P_{12}P_{21}$$

The sums yielding the allele frequencies are indicated by the diagonal connecting lines. The relation between D and the gametic frequencies can be verified by multiplying out $P_{11}P_{22} - P_{12}P_{21} = (p_1q_1 + D)(p_2q_2 + D) - (p_1q_2 - D)(p_2q_1 - D)$ and simplifying, using the facts that $p_1 + p_2 = 1$ and $q_1 + q_2 = 1$.

What linkage disequilibrium means biologically is that the frequency of the genotype $A_1 B_1 / A_2 B_2$ is not equal to the frequency of the genotype $A_1 B_2 / A_2 B_1$, even though each genotype is heterozygous for both loci. With random mating the frequencies of these genotypes are $P_{11}P_{22}$ and $P_{12}P_{21}$,

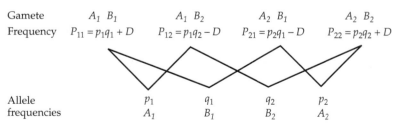

Gamete $\quad\quad A_1\ B_1 \quad\quad\quad A_1\ B_2 \quad\quad\quad A_2\ B_1 \quad\quad\quad A_2\ B_2$

Frequency $\quad P_{11} = p_1q_1 + D \quad P_{12} = p_1q_2 - D \quad P_{21} = p_2q_1 - D \quad P_{22} = p_2q_2 + D$

Allele $\quad\quad\quad\quad\quad\quad p_1 \quad\quad\quad\quad q_1 \quad\quad\quad\quad q_2 \quad\quad\quad\quad p_2$

frequencies $\quad\quad\quad\quad A_1 \quad\quad\quad\quad B_1 \quad\quad\quad\quad B_2 \quad\quad\quad\quad A_2$

$P_{11} + P_{12} + P_{21} + P_{22} = 1$ as well as $p_1 + p_2 = 1$ and $q_1 + q_2 = 1$

Figure 1.14   Two-locus gametes and their frequencies.

respectively. They are equal if and only if $D = 0$. On the other hand, each individual locus is in HWE. Random union of the gametes in Figure 1.14 results in genotype frequencies given by successive terms in the expansion of

$$(P_{11} A_1 B_1 + P_{12} A_1 B_2 + P_{21} A_2 B_1 + P_{22} A_2 B_2)^2$$

HWE for each locus can be confirmed directly from this binomial. For example, the frequency of A_1A_1 is $(P_{11} + P_{12})^2 = p_1^2$ (the squared term of HWE), that of A_1A_2 is $2(P_{11} + P_{12})(P_{21} + P_{22}) = 2p_1p_2$ (the cross-product term of HWE), and so forth for the other genotypes.

Linkage equilibrium between genes is eventually attained under random mating and the assumptions for HWE. The attainment of LE is gradual, however, in contrast to the attainment of HWE for the alleles of a single gene, which typically requires only one or a small number of generations. The rate of approach to LE again depends on the double heterozygous genotypes $A_1 B_1 / A_2 B_2$ and $A_1 B_2 / A_2 B_1$ and, in particular, on the relative frequencies of the several types of gametes that they can produce. If the genes are in different chromosomes, or are very far apart in the same chromosome, then each double heterozygous genotype produces the four possible types of gametes $(A_1 B_1, A_2 B_2, A_1 B_2,$ and $A_2 B_1)$ in equal frequencies. In this case, as will be shown below, the value of D decreases by half in each successive generation.

When two genes are sufficiently close together in the same chromosome, they are said to be **linked genes,** and the frequencies of the different types of gametes depends on the distance between the genes. This is because each chromosome aligns side-by-side with its partner chromosome during the formation of gametes and can undergo a sort of breakage and reunion resulting in an exchange of parts between the partner chromosomes. The distance between the genes determines the likelihood of an exchange. The result of genetic exchange is called *recombination,* and it is measured by the **frequency of recombination,** r, defined as the proportion of gametes that carry a combination of alleles not present in either parental chromosome. Recombination is illustrated in Figure 1.15. Part A pertains to the parental double heterozygote $A_1 B_1 / A_2 B_2$, in which the slash separates the alleles present in homologous chromosomes. In this case the nonrecombinant (parental) chromosomes are $A_1 B_1$ and $A_2 B_2$, whereas the recombinant chromosomes are $A_1 B_2$ and $A_2 B_1$. The frequency of each type of chromosome is indicated, and the overall frequency of recombinant chromosomes is r. The situation is the same in the double heterozygote $A_1 B_2 / A_2 B_1$ shown in part B, except that the nonrecombinant and recombinant chromosomes are reversed. The minimum frequency of recombination is 0, when the genes are completely linked. The maximum frequency of recombination is 0.5, when the genes are either far apart in the same chromosome or present in different chromosomes. The reason for the maximum at 0.5 is that the more exchanges there

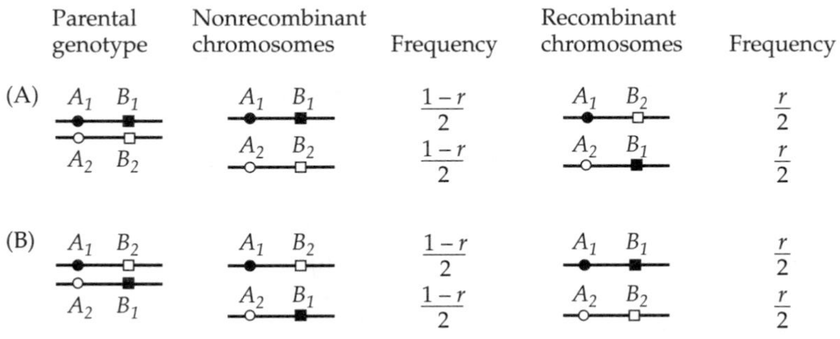

Parental genotype	Nonrecombinant chromosomes	Frequency	Recombinant chromosomes	Frequency

Figure 1.15   Consequences of recombination in two types of double heterozygotes (A and B).

are, the more likely that one exchange will undo the result of a previous exchange, until eventually the alleles in the parental chromosomes become randomly combined in the gametes. This means that the gametic types are equally frequent; in other words $r = 0.5$.

The recombination fraction between genes determines the rate of approach to linkage equilibrium. To see how, consider a chromosome carrying $A_1 B_1$. This chromosome could have only two possible origins relative to the chromosomes in the previous generation. It could be a recombinant chromosome with probability r, or it could be a nonrecombinant chromosome with probability $1 - r$. If it is a recombinant chromosome, it must have come from a parent of genotype $A_1 - / - B_1$, where in this case the dash means that it is unnecessary to specify the particular allele at the locus. The frequency of this genotype is p_1q_1 because of the random mating. On the other hand, if it is a nonrecombinant chromosome, then its progenitor chromosome in the previous generation must also have been $A_1 B_1$, and the frequency of this type of chromosome is P_{11}. Putting these possibilities together, and letting P_{11}' be the frequency of $A_1 B_1$ chromosomes in the present generation,

$$P_{11}' = rp_1q_1 + (1 - r)P_{11}$$

Subtracting p_1q_1 from both sides leads to

$$P_{11}' - p_1q_1 = (1 - r)(P_{11} - p_1q_1)$$

However, the parameterizations in Figure 1.14 indicate that $P_{11} - p_1q_1 = D$. This means that in each generation D is only $1 - r$ times as large as in the previous generation, Hence, letting D_t be the value of D in generation t,

$$D_t = D_{t-1}(1 - r) = D_{t-2}(1 - r)^2 = D_{t-3}(1 - r)^3 = \cdots = D_0(1 - r)^t \quad (1.10)$$

where D_0 is the value of D in the initial or founder population. Because $1 - r < 1$, then $(1 - r)^t$ goes to zero as t becomes large, but how rapidly it does depends on r—the closer to zero, the slower the rate. Recall here that $r = 0.5$ corresponds either to genes far apart in the same chromosome or to genes in different chromosomes; it implies that the disequilibrium decreases by half in each generation.

One limitation of D as a measure of linkage disequilibrium is that its possible values depend on the allele frequencies; hence it is difficult to compare values of D from one pair of loci to the next. For this reason D is sometimes expressed as a percentage of its minimum value (if D is negative) or as a percentage of its maximum value (if D is positive), which restricts the range from 0 to 100 percent. The minimum and maximum values of D can be deduced from Figure 1.14 by noting that all four gametic frequencies must be nonzero. This implies that

$$D_{min} = \text{the larger of } -p_1q_1 \text{ and } -p_2q_2$$

$$D_{max} = \text{the smaller of } p_1q_2 \text{ and } p_2q_1 \tag{1.11}$$

Another measure of linkage disequilibrium having a predefined range is

$$\rho = \frac{D}{\sqrt{p_1 p_2 q_1 q_2}} \tag{1.12}$$

which is the correlation between the A and B alleles present in gametes and can range from -1 to $+1$. This measure has the convenient feature that the X^2 value for goodness of fit to the hypothesis that $D = 0$ is given by

$$X^2 = \rho^2 n \tag{1.13}$$

where n is the number of chromosomes in the sample. The statistical significance of this X^2 value is obtained as before from Figure 1.11 for the curve with one degree of freedom. A statistically significant value of X^2 leads to rejection of the hypothesis that $D = 0$, but rather large sample sizes are usually required unless the disequilibrium is extreme (Brown 1975; Zapata and Alvarez 1993; Valdes and Thomson 1997).

An example of LD is found in the genes for the human MN and Ss blood groups, which are determined by closely linked genes in chromosome 4. Neither of these blood groups is important in clinical medicine, but they typify the classical blood group polymorphisms that were the mainstay of human population genetics until the advent of protein and especially DNA analysis. We will identify the alleles M and N with A_1 and A_2 in Figure 1.14

and the alleles S and s with B_1 and B_2. A sample of 1000 British people yielded the following numbers of each chromosome type:

$M\,S$ 474 $\qquad$ $M\,s$ 611 $\qquad$ $N\,S$ 142 $\qquad$ $N\,s$ 773

From these data the allele frequencies are $\langle p_1 \rangle$ = 0.542, $\langle p_2 \rangle$ = 0.458, $\langle q_1 \rangle$ = 0.308, and $\langle q_2 \rangle$ = 0.692. The estimated value of $\langle D \rangle$ = (474 × 773 − 611 × 142)/(2000)^2 = 0.0699. Is this value significantly different from 0? To find out, we calculate $\rho = D/\sqrt{p_1 p_2 q_1 q_2} = 0.3040$, and then $X^2 = \rho^2 \times n = 184.78$ with 1 degree of freedom. This value is off the chart in Figure 1.11, so we may conclude that P is very much less than 0.0001. This means that chance alone would produce a fit as poor or poorer substantially less than one time in 10,000, so the hypothesis that the loci are in linkage equilibrium can confidently be rejected.

Although D is statistically highly significant, is it large value or a small value? To assess this issue, we calculate D as a percentage of its maximum possible value. (If D were negative, we would use the minimum possible value.) The maximum is given by the smaller of $p_1 q_2$ = 0.375 and $p_2 q_1$ = 0.141, and so D_{max} = 0.141. The ratio D/D_{max} = 0.496; hence we can say that D is about 50% of its maximum possible value, given the allele frequencies. It is also instructive to calculate the expected number of each of the chromosome types when D = 0, namely,

$M\,S$ 334.2 $\qquad$ $M\,s$ 750.8 $\qquad$ $N\,S$ 281.8 $\qquad$ $N\,s$ 633.2

Direct calculation confirms the X^2 value, except for a small round-off error.

For most pairs of genes in natural populations, D is usually zero or close to zero (indicating linkage equilibrium) unless the genes are very tightly linked (separated by less than about 100 kb). One exception applies to genes present in or near chromosomal inversions. These, as the name implies, have a segment reversed relative to the normal orientation. Topological problems prevent recovery of genetic exchanges within the inverted segment between a normal and an inversion-bearing chromosome, and the net effect is to create tight linkage between all genes in the inversion, so that effectively r = 0. Inversions and the consequent linkage disequilibrium are very frequent in certain a species of *Drosophila*, especially *D. pseudoobscura* and its relatives.

Another exception to the rule of generally absent or small LD in natural populations pertains to plants that regularly undergo self-fertilization. As we shall see in the next section, close inbreeding severely reduces the frequency of heterozygous genotypes, and hence severely reduces the frequency of the double heterozygous genotypes wherein recombination helps dissipate the linkage disequilibrium. Close inbreeding therefore greatly decreases the opportunities for recombination, thereby effectively decreasing r. LD can also be caused by linkage disequilibrium in the founding population that has not yet had time to decay through recombination. Another possible cause of LD is admixture of populations with differing gametic frequencies. Just as an admixture of two HWE subpopulations differing in allele frequencies can result in a departure from HWE in the direction of too few heterozygotes in the mixed population, an admixture of two LE populations differing in gametic frequencies can result in LD in the mixed population. The LD decays gradually, even for pairs of genes that are unlinked. Yet another possible reason for the persistence of LD is natural selection if it favors certain genotypes with sufficient intensity to overcome the natural tendency for LD to gradually disappear.

INBREEDING

Mating between relatives is **inbreeding.** The main effect of inbreeding is to increase the frequency of homozygous genotypes in a population, relative to the frequency that would be expected with random mating. Unlike random mating, which may affect some genes but not others, inbreeding affects all genes in the genome. In human populations, the closest degree of inbreeding that commonly occurs in most societies is first-cousin mating, but many plants regularly undergo such close inbreeding as self-fertilization.

Genotype Frequencies with Inbreeding

The principal effect of inbreeding in a population is to increase the frequency of homozygous genotypes at the expense of the frequency of heterozygous genotypes. This effect can most easily be seen in the case of repeated self-fertilization. Consider a self-fertilizing population of plants that consists of 25% *AA*, 50% *Aa*, and 25% *aa* genotypes. These genotype frequencies are in HWE. Because each plant undergoes self-fertilization, the *AA* and *aa* genotypes produce only *AA* and *aa* offspring, respectively, whereas the *Aa* genotypes produce 1/4 *AA*, 1/2 *Aa*, and 1/4 *aa* offspring. After one generation of self-fertilization, therefore, the genotype frequencies of *AA*, *Aa*, and *aa* are

$$\left[\frac{1}{4}(1)+\frac{1}{2}\left(\frac{1}{4}\right)\right] AA \qquad \left[\left(\frac{1}{2}\right)\left(\frac{1}{2}\right)\right] Aa \qquad \left[\frac{1}{4}(1)+\frac{1}{2}\left(\frac{1}{4}\right)\right] aa \qquad (1.14)$$

or

$$(3/8)\ AA \qquad\qquad (2/8)\ Aa \qquad\qquad (3/8)\ aa$$

These genotype frequencies are no longer in HWE because there is a deficiency of heterozygous genotypes and an excess of homozygous genotypes.

The effects of inbreeding can be made quantitative in terms of the reduction in heterozygosity. This means that we can measure the amount of inbreeding by comparing the actual proportion of heterozygous genotypes in the population with the proportion of heterozygous genotypes that would occur with random mating. To be concrete, consider a locus with two alleles, A and a, at respective frequencies p and q (with $p + q = 1$). Suppose that the actual frequency of heterozygous genotypes in a population at the present time is denoted H. If the population were in HWE for the gene, the frequency of heterozygous genotypes would be $2pq$. We will denote this baseline value as H_0, hence $H_0 = 2pq$. The effects of inbreeding can be defined in terms of the quantity $(H_0 - H)/H_0$, which is usually denoted in population genetics by the symbol F and called the **inbreeding coefficient.** Thus

$$F = \frac{(H_0 - H)}{H_0} \tag{1.15}$$

Biologically, F measures the reduction in heterozygosity, measured as a fraction relative to that expected in a random-mating population with the same allele frequencies. Because $H_0 = 2pq$, the actual frequency of heterozygous genotypes in the inbred population can be written in terms of F as

$$H = H_0 - H_0 F = 2pq - 2pqF \tag{1.16}$$

The frequency of AA homozygous genotypes in the inbred population can also be expressed in terms of F. Suppose that the proportion of AA genotypes is actually D. Because the allele frequency of A is p, we must have, by Equation 1.1, that $D + H/2 = p$. But $H = 2pq - 2pqF$. Therefore,

$$D = p - \frac{(2pq - 2pqF)}{2} = p^2 + pqF \tag{1.17}$$

Likewise the frequency of aa genotypes R is

$$R = q - \frac{(2pq - 2pqF)}{2} = q^2 + pqF \tag{1.18}$$

A little algebraic manipulation of Equations 1.16–1.18 enables the genotype frequencies with inbreeding to be written in the alternative form:

$$AA: \quad p^2(1-F)+pF$$
$$Aa: \quad 2pq(1-F) \qquad\qquad (1.19)$$
$$aa: \quad q^2(1-F)+qF$$

This formulation shows that the genotype frequencies equal the HWE frequencies multiplied by the factor $1 - F$, plus a correction term for the homozygous genotypes multiplied by the factor F. When $F = 0$ (no inbreeding), the genotype frequencies are the HWE. When $F = 1$ (complete inbreeding), the population consists entirely of AA and aa homozygous genotypes in the frequencies p and q, respectively.

The reorganization of genotype frequencies with inbreeding is illustrated graphically in Figure 1.16. To obtain the genotype frequencies, multiply $1 - F$ times the HWE frequencies in the rectangles on the left, and add F times the allele frequencies in the rectangles on the right. This calculation corresponds to Equation 1.19. The main effect of inbreeding is that some heterozygous genotypes disappear from the population and are replaced with homozygous genotypes. This transfer is indicated by the arrows. Why it occurs is explained in the next section.

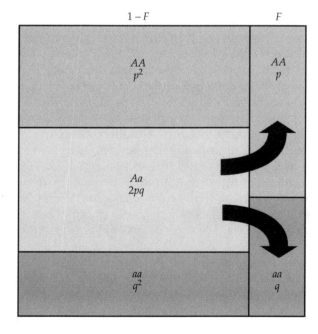

Figure 1.16 Genotype frequencies with inbreeding. The effect of inbreeding is to eliminate some heterozygous genotypes and replace them with homozygous genotypes.

The Inbreeding Coefficient

The inbreeding coefficient F has an alternative interpretation in terms of probability. For any pair of alleles present in a single inbred individual, the alleles are called **identical by descent** (**IBD**) if they are both derived by DNA replication of a single allele present in some ancestral population. The probability interpretation of F is that, for a pair of alleles present in an inbred individual, *F is the probability that the alleles are IBD*. When expressed in terms of the probability of IBD, the inbreeding coefficient is clearly a relative concept because we have not specified the arbitrary ancestral population in which we define all pairs of alleles as being *not* IBD. In other words, we regard the ancestral population as having no inbreeding ($F = 0$). Relative to this ancestral population, the inbreeding coefficient of an individual in the present population is the probability that the two alleles at a locus in the individual arose by replication of a single allele more recently than the time at which the ancestral population existed. The ancestral population need not be remote in time from the present one. The ancestral population typically refers to the population existing just a few generations previous to the present one, and F in the present population then measures inbreeding that has occurred in the span of these few generations. Because the span of time involved is usually short, the possibility of mutation can safely be ignored.

If the two alleles in an inbred individual are IBD, the genotype at the locus is said to be **autozygous.** If they are not IBD, the genotype is said to be **allozygous.** The distinction is important because an autozygous genotype must be homozygous, since the alleles are IBD and we assume no mutation. On the other hand, allozygous genotypes can be either homozygous or heterozygous. Figure 1.17 illustrates how the concepts of autozygosity and allozygosity are related to those of homozygosity and heterozygosity. The essential point is that two alleles can be chemically identical (in terms of having the same sequence of nucleotides along the DNA) without being identical by descent; the concept of identity by descent pertains to the ancestral origin of an allele and not to its chemical makeup. Although, as shown in Figure 1.17, two distinct alleles that are chemically identical (two A_1s or two A_2s, for example) may come together in an individual and thereby make the individual homozygous, the alleles in the ancestral population are, by definition, not identical by descent, so the individual is allozygous. Similarly, although a heterozygous individual must be allozygous (ignoring mutation), a homozygous individual may be either autozygous or allozygous (see Figure 1.17).

To show that the inbreeding coefficient defined in terms of IBD is equivalent to that defined in Equation 1.15 in terms of heterozygosity, we need only consider the implications of the probability definition for an entire population. Imagine, therefore, a population in which individuals have average inbreeding coefficient F. Focus on one individual, and consider the alleles of

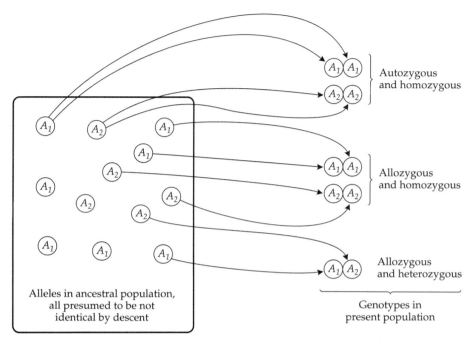

Figure 1.17 In an autozygous individual, homologous alleles are derived from replication of a single DNA sequence in an ancestor and are therefore IBD. In an allozygous individual, homologous alleles are not IBD.

any gene in the individual. Either of two things must be true: the alleles must be either allozygous (probability $1 - F$) or autozygous (probability F). If the alleles are allozygous, then the probability that the individual has any particular genotype is the same as the probability of that genotype with HWE, because, by chance, the inbreeding has not affected this particular gene. On the other hand, if the alleles are autozygous, then the individual must be homozygous, and the probability that the individual is homozygous for any particular allele is equal to the frequency of that allele in the population as a whole. (Because the alleles in question are autozygous, knowing which allele is present in one chromosome immediately tells you that an identical allele is in the homologous chromosome.) In symbols, the probability that an individual has genotype AA is $p^2(1 - F)$ [when the alleles are allozygous] + pF [when the alleles are autozygous]. Similarly, the probability that the individual has genotype aa is $q^2(1 - F) + qF$. Heterozygous Aa genotypes then occur in the frequency $2pq(1 - F)$, since alleles that are heterozygous must be allozygous. The genotype frequencies with inbreeding are summarized in Table 1.3. Note that the genotype frequencies are exactly the same as those

Table 1.3 Genotype frequencies with inbreeding

Genotype	With inbreeding coefficient F			With $F = 0$ (random mating)	With $F = 1$ (complete inbreeding)
AA	$p^2(1 - F)$	$+$	pF	p^2	p
Aa	$2pq(1 - F)$			$2pq$	0
aa	$q^2(1 - F)$	$+$	qF	q^2	q
	Allozygous genes		Autozygous genes		

given in Equation 1.19; the similarity shows that the IBD definition of F and the heterozygosity definition of F, though superficially quite different, are actually equivalent.

Inbreeding Depression

In species that usually breed by outcrossing with unrelated individuals, close inbreeding is generally harmful. These harmful effects are referred to as **inbreeding depression** (Charlesworth and Charlesworth 1987; Frankham 1995). Figure 1.18 is an example of inbreeding depression in yield of corn. The harmful effects are seen most dramatically when inbreeding is complete or nearly complete. Although nearly complete inbreeding can be approached in most species by many generations of brother–sister mating, autozygosity of whole chromosomes can easily be accomplished in *Drosophila* through the use of special inversion-bearing chromosomes called *balancer chromosomes* that eliminate all recombination. The results of a typical experiment in which whole chromosomes from natural populations are rendered IBD are shown in Figure 1.19. The crosses are carried out in such a way as either to make a single natural chromosome homozygous (also IBD) or to make two different natural chromosomes heterozygous (allozygous). The distribution of relative survivorship (viability) of the resulting genotypes is shown. A relative viability of 1.0 corresponds to equality with a standard genotype present in each culture bottle. It is evident that the homozygous genotypes (gray outline) are relatively poor in viability. In fact, about 12.5% of the homozygous chromosomes are lethal. Moreover, among the homozygotes that have viabilities within the normal range of heterozygotes (black outline), virtually all can be shown to have reduced fertility (Sved 1975; Simmons and Crow 1977). The harmful inbreeding effects are due mainly to rare recessive alleles that are severely detrimental when homozygous. The inference that each harm-

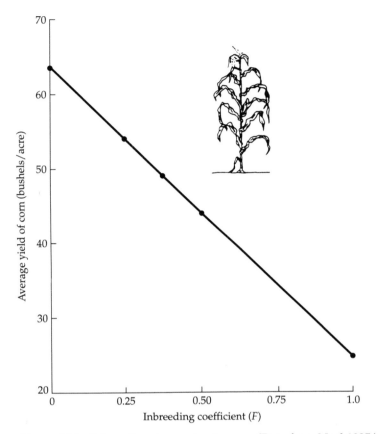

Figure 1.18   Inbreeding depression in corn. (Data from Neal 1935.)

ful recessive allele is rare is supported by the small proportion of lethal or near-lethal heterozygous chromosomes.

As in other outcrossing organisms, inbreeding in humans is generally harmful, but the effects are difficult to measure because the degree of inbreeding is usually quite small and the effects may also vary from population to population. The effects are again due largely to the increased homozygosity of rare recessive alleles. They are observed most dramatically in the increased frequency of inherited diseases due to harmful recessive alleles among the children of matings between first cousins or other degrees of familial relationship. For a rare deleterious recessive allele, the frequency of homozygous recessives among the children of first-cousin matings (for which $F = 1/16$) is given by $q^2(1 - 1/16) + q(1/16)$. On the other hand, with HWE the frequency of recessive homozygotes is q^2. Therefore, the risk of an

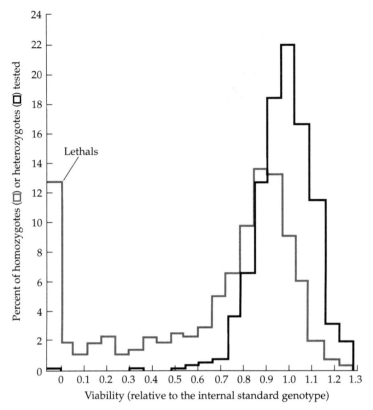

Figure 1.19   Viability distributions of homozygous chromosomes (gray outline) and heterozygous chromosomes (black outline) of second chromosomes extracted from a natural population of *Drosophila pseudoobscura*. There were 1063 homozygous and 1034 heterozygous genotypes tested.

affected offspring from a first-cousin mating, relative to that from a mating of nonrelatives, is given by

$$\text{Relative risk} = \frac{q^2\left(1-\frac{1}{16}\right)+q\left(\frac{1}{16}\right)}{q^2} = 0.9375 + \frac{0.0625}{q} \tag{1.20}$$

For example, when $q = 0.01$, the relative risk is approximately 7, which means that a first-cousin mating has about a 700% greater risk of producing a homozygous recessive child than a mating between nonrelatives. This is clearly a large inbreeding effect, and the rarer the frequency of the deleterious recessive allele, the greater the relative risk.

One of the dramatic successes of plant breeding has come from crossing different inbred lines of corn to produce high-yielding hybrid varieties. Although inbred lines are inferior due to inbreeding depression (Figure 1.18), different inbred lines are not likely to become homozygous for exactly the same set of deleterious recessive genes. Therefore, when different inbred lines are crossed to produce a hybrid, the hybrid is heterozygous for most or all of the deleterious recessives fixed in the inbreds. Alleles favoring high yield in corn are generally dominant, so the hybrid has a much higher yield than either inbred parent. The hybrid plants are also quite uniform because they all have the same genotype. The phenomenon of enhanced hybrid performance is called hybrid vigor or heterosis. In practice, inbred lines are crossed in many combinations to identify those that produce the best hybrids. Yields of hybrid corn are typically 15% to 35% greater than yields of outcrossing varieties. Hybrids account for virtually 100% of all corn cultivated in the United States today, as compared with less than 1% in 1933 (Sprague 1978).

Calculation of the Inbreeding Coefficient from Pedigrees

Computation of F from a pedigree is simplified by drawing the pedigree in the form shown in Figure 1.20A, where the lines represent gametes contributed by parents to their offspring. The same pedigree is shown in conventional form in Figure 1.20B. The individuals in gray in B are not represented in A because they have no ancestors in common and therefore do not contribute to the inbreeding of individual I. The inbreeding coefficient F_I of individual I is the probability that I carries alleles of an arbitrarily chosen gene that are IBD. The first step in calculating F_I is to identify all common ancestors in the pedigree, because an ancestral allele could become IBD in I only if it were inherited through both of I's parents from a common ancestor. In this case there is only one common ancestor, labeled A. The next step in calculating F_I, which is carried out for each common ancestor in turn, is

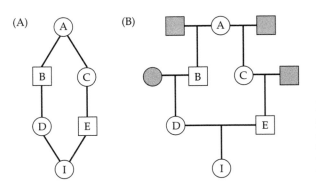

Figure 1.20 (A) Pedigree diagram to facilitate calculation of the inbreeding coefficient. (B) Conventional representation of the same pedigree.

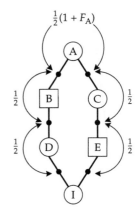

Figure 1.21   Identity loops for the pedigree in Figure 1.20. Each number is the probability of IBD for the alleles indicated.

to trace all the paths of gametes that lead from one of I's parents back to the common ancestor and then down again to the other parent of I. These paths are the paths along which an allele in a common ancestor could become IBD in the individual I. In Figure 1.20A, there is only one such path: DB$\underline{A}$CE. The common ancestor has been underlined for bookkeeping purposes, an especially useful procedure in complex pedigrees.

The third step in calculating F_I is to calculate the probability of IBD in I due to each of the paths in turn. For the path DBACE, the reasoning involved is illustrated in Figure 1.21. Here the black dots represent alleles transmitted along the gametic paths, and the number associated with each loop is the probability of identity by descent of the alleles indicated. For all individuals except the common ancestor, the probability is 1/2, because with Mendelian segregation the probability that a particular allele present in a parent is transmitted to a specified offspring is 1/2. To understand why $(1/2)(1 + F_I)$ is the probability associated with the loop around the common ancestor, denote the alleles in the common ancestor as α_1 and α_2. (The Greek symbols are used to avoid confusion with conventional allele symbols designating chemical types of alleles, such as A and a for dominant and recessive, respectively.) The pair of gametes contributed by individual A could contain $\alpha_1 \alpha_1$, $\alpha_2 \alpha_2$, $\alpha_1 \alpha_2$, or $\alpha_2 \alpha_1$, each with a probability of 1/4 because of Mendelian segregation. In the first two cases, the alleles are clearly IBD; in the second two cases, the alleles are IBD only if α_1 and α_2 are IBD, and α_1 and α_2 are IBD only if individual A is autozygous, which has probability F_A, the inbreeding coefficient of A. Altogether, the required probability for the loop around individual A is $1/4 + 1/4 + (1/4)F_A + (1/4)F_A = 1/2 + (1/2)F_A = (1/2)(1 + F_A)$. Because each of the loops in Figure 1.21 is independent of the others, the total probability of autozygosity in individual I due to this path is $1/2 \times 1/2$

× (1/2)(1 + F_A) × 1/2 × 1/2, or $(1/2)^5(1 + F_A)$. Note that the exponent on the 1/2 is equal to the number of individuals in the path. In general, if a path through a common ancestor A contains i individuals, the probability of autozygosity due to that path is

$$\left(\frac{1}{2}\right)^i (1+F_A)$$

Thus, the inbreeding coefficient of individual I in Figure 1.20A, assuming $F_A = 0$, equals $(1/2)^5 = 1/32$.

In more complex pedigrees, there is more than a single path. The paths are mutually exclusive, because if the alleles are IBD due to being inherited along one path, they cannot at the same time be IBD due to being inherited along a different path. Therefore, the total inbreeding coefficient is the sum of the probability of IBD due to each separate path. The whole procedure for calculating F is summarized in an example involving a first-cousin mating in Figure 1.22. Here there are two common ancestors (A and B) and two paths (one each through A and B). The total inbreeding coefficient of I is the sum of the two mutually exclusive contributions shown in Figure 1.22. If A and B are noninbred, then $F_A = F_B = 0$, and $F_I = (1/2)^5 + (1/2)^5 = 1/16$. This is the probability that an arbitrarily chosen pair of alleles in I are IBD. Alternatively, F_I can be interpreted as the average proportion of all pairs of alleles in I that are IBD.

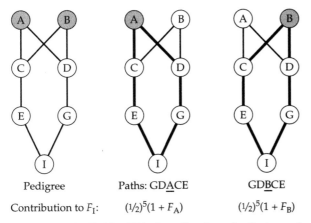

Pedigree	Paths: GD<u>A</u>CE	GD<u>B</u>CE
Contribution to F_I:	$(½)^5(1 + F_A)$	$(½)^5(1 + F_B)$

Figure 1.22 Individual I is the offspring of a first-cousin mating. Two paths (heavy lines) through common ancestors A and B are shown at the right.

In general, for any gene except those in the X and Y chromosomes, the formula for calculating the inbreeding coefficient F_I of an individual I is

$$F_I = \sum_{all\,paths} \left(\frac{1}{2}\right)^i \left(1 + F_A\right) \tag{1.21}$$

where the summation is over all of the possible paths through all common ancestors, i is the number of individuals in each path, and A is the common ancestor in each path.

Regular Systems of Mating

Some domesticated animals and cultivated plants are propagated by a regular system of mating, such as repeated self-fertilization, sib mating, or backcrossing to a standard strain. It is then of interest to know how the inbreeding coefficient increases with time. The reasoning involved for the special case of self-fertilization has already been discussed in connection with Equation 1.14, from which it is clear that the inbreeding coefficient in generation t (F_t) is related to that in generation $t-1$ (F_{t-1}) by

$$1 - F_t = \frac{1}{2}\left(1 - F_{t-1}\right) \tag{1.22}$$

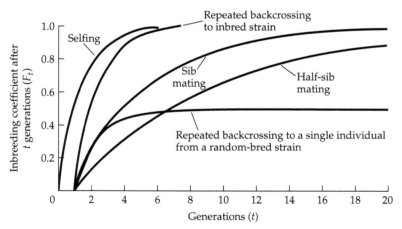

Figure 1.23 Theoretical increase in the inbreeding coefficient for various regular systems of mating.

because the frequency of heterozygous genotypes decreases by a factor of $1/2$ in each generation. It follows that for repeated self-fertilization

$$1 - F_t = \left(\frac{1}{2}\right)^t (1 - F_0)$$ (1.23)

How the inbreeding coefficient increases under various regular systems of mating is shown in Figure 1.23. In most cases, F increases gradually with time. One important general principle is that, no matter what the value of F in a population at any time, and no matter how this value of F was created, the inbreeding coefficient immediately returns to 0 (no inbreeding) with just one generation of outcrossing to an unrelated strain.

FURTHER READINGS

Avise, J. C. 1994. *Molecular Markers, Natural History and Evolution*. Chapman and Hall, New York.

Cavalli-Sforza, L. L. 1996. *History and Geography of Human Genes*. Princeton University Press, Princeton, NJ.

Cavalli-Sforza, L. L. and W. F. Bodmer. 1971. *The Genetics of Human Populations*. W. H. Freeman and Co., San Francisco.

Chakravarti, A. 1984. *Human Population Genetics*. Van Nostrand Reinhold, New York.

De Jong, G. 1988. *Population Genetics and Evolution*. Springer-Verlag, Berlin.

Edwards, A. W. F. 1977. *Foundations of Mathematical Genetics*. Cambridge University Press, Cambridge.

Evett, I. W. and B. S. Weir. 1998. *Interpreting DNA Evidence: Statistical Genetics for Forensic Scientists*. Sinauer Associates, Sunderland, MA.

Feldman, M. W. and F. B. Christiansen. 1986. *Population Genetics*. Blackwell Scientific Publications, Palo Alto, CA.

Hartl, D. L. and A. C. Clark. 1997. *Principles of Population Genetics*, 3rd Ed. Sinauer Associates, Sunderland, MA.

Li, C. C. 1976. *First Course in Population Genetics*. Boxwood Press, Pacific Grove, CA.

Provine, W. B. 1987. *The Origins of Theoretical Population Genetics*. University of Chicago Press, Chicago.

Wallace, B. 1981. *Basic Population Genetics*. Columbia University Press, New York.

Wright, S. 1968–1978. *Evolution and the Genetics of Populations*. Vol. 1, 1968: *Genetic and Biometric Foundations*. Vol. 2, 1969: *The Theory of Gene Frequencies*. Vol. 3, 1977: *Experimental Results and Evolutionary Deductions*. Vol. 4, 1978: *Variability within and among Natural Populations*. University of Chicago Press, Chicago.

PROBLEMS

The accompanying figure has diagrams of gels showing DNA polymorphisms detected in population samples. Above each lane is the number of individuals in the sample with the banding pattern illustrated. Some of the problems below are based on interpreting such gels.

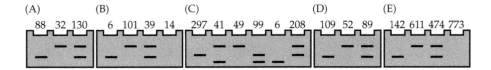

(A)   (B)   (C)   (D)   (E)

88   32   130   6   101   39   14   297   41   49   99   6   208   109   52   89   142   611   474   773

1.1 Gel A shows an RFLP pattern due to the segregation of homologous DNA fragments differing in the positions of certain restriction sites. Estimate the allele frequencies and carry out a chi-square test for HWE.

1.2 Gel B shows the banding patterns obtained for two RAPD fragments.

    a. Estimate the frequency q of the null allele of each RAPD assuming HWE.

    b. Estimate the percentage of individuals showing each RAPD band that are heterozygous for the null allele.

    c. Calculate the frequency q for which half the individuals with the dominant phenotype are heterozygous.

    d. Although HWE cannot be tested for either RAPD individually, there is one degree of freedom allowing a chi-square test for independence between the RAPDs. Perform this test.

1.3 Gel C shows a three-allele RFLP. Estimate the allele frequencies and carry out a chi-square test for goodness of fit to HWE.

1.4 Phenylketonuria is a severe form of mental retardation caused by a homozygous recessive allele. The condition affects about 1 in 10,000 newborn Caucasians. Estimate the frequency of heterozygotes for this allele assuming HWE.

1.5 Industrial melanism refers to the dark pigmentation that evolved in some insects, giving them protective coloration on vegetation darkened by soot in heavily industrialized areas prior to the requirement for smokestack filtration. In one heavily polluted area near Birmingham, England in 1956, 87% of moths of the species *Biston betularia* had black bodies due to the presence of a dominant gene for melanism (Kettlewell 1956). Estimate the frequency of the dominant allele in this population and the frequency of melanics that are heterozygous.

1.6 Suppose that both sexes in a diploid population have genotype frequencies of *AA, Aa,* and *aa* equal to *P, Q,* and *R,* where $P + Q + R = 1$. Show that the frequencies are in HWE if and only if $Q^2 = 4PR$.

1.7 In a sample of 1617 Spanish Basques, the numbers of A, B, O, and AB blood types observed were 724, 110, 763, and 20, respectively. These blood groups are due to three alleles, I^A, I^B, and I^O, with $I^A I^A$ and $I^A I^O$ having blood group A; $I^B I^B$ and $I^B I^O$ having blood group B; $I^A I^B$ having blood group AB; and $I^O I^O$ having blood group O. The best estimates of allele frequency in the Basque sample are 0.2661 for I^A, 0.0411 for I^B, and 0.6928 for I^O. Calculate the expected numbers of the four blood group phenotypes and carry out a chi-square test for HWE.

1.8   Among many aboriginal American Indian tribes, the allele frequency of I^B in the ABO blood groups is extremely low. For example, in one sample of 600 Papago Indians from Arizona, there were 37 A and 563 O blood types. What are the best estimates of the allele frequencies of I^A, I^B, and I^O in this population, and what are the expected genotype frequencies assuming HWE?

1.9   Four RFLP alleles in a random mating population have frequencies 4/10, 3/10, 2/10, and 1/10. What are the expected genotype frequencies? What is the probability of a matching DNA type for this RFLP between two unrelated individuals?

1.10   Gel D shows a two-allele RFLP. Estimate the allele frequencies and test for goodness of fit to HWE. If the sample were from a plant that could undergo self-fertilization, what would you conclude about whether or not self-fertilization actually takes place? If there is evidence of inbreeding, estimate F.

1.11   A RAPD marker in human populations cosegregates with the X-linked Xga blood group polymorphism and is probably in the same gene. A sample of 2082 people yielded 967 females and 667 males with the RAPD band, and 102 females and 346 males lacking the band. Using the average allele frequencies in the two sexes, calculate the expected numbers in the four phenotypic classes assuming HWE, and carry out a chi-square test for goodness of fit. (The number of degrees of freedom in this case is 1, even though only the null allele frequency needs to be calculated, because the total numbers of females and males are both fixed.)

1.12   Gel E shows two RAPD markers in samples of the gametophyte (haploid) phase of the life cycle of a liverwort.

   a. Estimate the gametic frequencies and test for linkage disequilibrium.
   b. If there is significant LD, calculate the amount of disequilibrium relative to its maximum or minimum value.

1.13   Mukai et al. (1974) captured 660 fertilized females of *D. melanogaster* from a natural population in Raleigh, NC. These females were used to found a large experimental population. After about five months (10 generations), 489 third chromosomes in the population were analyzed for allozymes coding for the enzymes esterase-6 (alleles $E6^F$ and $E6^S$), esterase-C (alleles EC^F and EC^S), and octanol dehydrogenase (alleles Odh^F and Odh^S). The results were as follows:

$E6^F$	EC^F	Odh^F	152	$E6^S$	EC^F	Odh^F	264
$E6^F$	EC^F	Odh^S	7	$E6^S$	EC^F	Odh^S	13
$E6^F$	EC^S	Odh^F	15	$E6^S$	EC^S	Odh^F	29
$E6^F$	EC^S	Odh^S	1	$E6^S$	EC^S	Odh^S	8

The order of the genes in the third chromosome is known to be $E6$—EC—Odh. The recombination fraction between $E6$ and EC is 0.122, and that between EC and Odh is 0.002. (That is, $E6$ and EC are rather loosely linked, and EC and Odh are tightly linked.)

a. Carry out an analysis to determine whether there is significant linkage disequilibrium between any of the loci.

b. If there is linkage disequilibrium, express the value of D relative to its theoretical maximum or minimum.

1.14 Cross and Birley (1986) studied restriction site variation in the region of the alcohol dehydrogenase (Adh) gene of $D.$ *melanogaster* in a population descended from flies trapped at a Dutch fruit market in Groningen. The following data were obtained:

Adh^F	EcoRI$^+$	22	Adh^F	EcoRI$^-$	3
Adh^S	EcoRI$^+$	4	Adh^S	EcoRI$^-$	5

where Adh^F and Adh^S are the allozyme alleles of Adh and EcoRI$^+$ and EcoRI$^-$ indicate the presence of absence of an EcoRI restriction site 3.5 kb downstream from Adh.

a. Test for the presence of linkage disequilibrium.

b. If significant, express D relative to its theoretical maximum or minimum.

1.15 Three linked loci in a random mating population have the gametic frequencies shown in the accompanying table. Calculate the linkage disequilibrium D/D_{max} between each pair of genes. Explain why the results seem paradoxical.

$A\,B\,C$	1/4	$A\,b\,C$	1/4	$a\,B\,C$	0	$a\,b\,C$	0
$A\,B\,c$	0	$A\,b\,c$	0	$a\,B\,c$	1/4	$a\,b\,c$	1/4

1.16 The accompanying pedigree shows an individual I produced by a mating of "three-half" cousins. (The parents are full cousins from the left-hand side of the pedigree and half cousins from the right-hand side.) Calculate the inbreeding coefficient of I,

a. assuming that none of the common ancestors is inbred.

b. assuming that any of the common ancestors may be inbred.

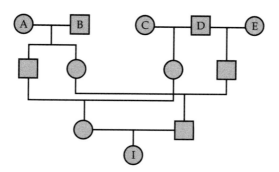

1.17 Calculate the risk of a homozygous recessive offspring from a mating of second cousins ($F = 1/64$) when the recessive allele frequency is $q = 0.01$, relative to the risk with random mating.

1.18 The pedigree shown below is for repeated backcrossing to the same male. Assume $F_0 = 0$.

    a. Calculate the inbreeding coefficients F_1, F_2, and F_3.

    b. Deduce a general formula for F_t.

    c. For an autosomal gene, would it matter whether the backcrossing were to the same male or to the same female? Why or why not?

    d. For an X-linked gene, would it matter whether the backcrossing were to the same male or to the same female? Why or why not?

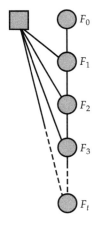

1.19 Segregation distortion is a phenomenon in which heterozygous genotypes do not produce the equal frequencies of complementary functional gametes expected from Mendelian segregation. A random-mating population has genotype frequencies of AA, Aa, and aa given by p^2, $2pq$, and q^2, respectively. A is a segregation distorter such that Aa heterozygotes produce a ratio $k : 1 - k$ of functional A-bearing and a-bearing gametes ($k > 1/2$).

    a. What are the expected allele frequencies among gametes?

    b. What are the expected genotype frequencies among zygotes in the next generation?

    c. Do HWE frequencies still hold in the zygotes? Explain why or why not.

1.20 This problem introduces Wahlund's principle of population admixture, which will be discussed further in Chapter 2. Consider a gene with two alleles A and a in two genetically isolated subpopulations A and B. Subpopulation A is fixed for allele A and subpopulation B is fixed for allele a. A "metapopulation" is produced that consists of an equal mixture of A and B.

    a. What are the genotype frequencies in the metapopulation?

    b. Are these consistent with HWE?

    c. What are the genotype frequencies in the metapopulation after one genera-
tion of random mating?

    d. How does the frequency of heterozygous genotypes in the original admixed
population differ from that after one generation of random mating?

    e. More generally, if A and B are both in HWE, but for different allele frequen-
cies, show that the average frequency of heterozygous genotypes in A and
B is always smaller than that expected after population fusion and random
mating of A and B.

**Solutions to the problems, worked out in full, can be found at the website
www.sinauer.com/hartl/html**

CHAPTER 2

The Causes of Evolution

The Hardy-Weinberg model ignores most of the complexities of actual populations. The allele frequencies in most natural populations are affected by mutation, migration, and natural selection. These processes can cause directional changes in allele frequency through time. Random fluctuations in allele frequency can also occur due purely to chance, because populations are not infinitely large and their sizes are rarely constant. All of these processes contribute to evolution because, in the widest sense, **evolution** can be defined as cumulative change in the genetic composition of a population. Some authors prefer a narrower definition that includes a stipulation that the genetic changes must be adaptive. One problem with a too-narrow definition is that there are many examples of genetic changes in populations, as well as genetic differences between species, whose adaptive significance is uncertain. In this chapter we consider the basic evolutionary processes of mutation, migration, natural selection, and random changes in allele frequency.

MUTATION

New genetic variation is created by changes in the genetic material; hence, mutation is the ultimate source of genetic variation. The term **mutation** is used here in a widest sense to mean all genetic changes, including nucleotide substitutions, insertions and deletions, changes in the genomic location of transposable genetic elements, and chromosome rearrangements. The creative role of mutation in evolution is exemplified by a gene in *D. melanogaster* that evolved in the 3.5 million years since the species split from its nearest relative, *D. simulans*. The gene codes for a novel dynein component present in the axoneme of the sperm tail, and its origin and evolution required a gene duplication, fusion of the duplicated genes via three deletions, two additional insertions/deletions (including one that created a novel splice junction), 11 nucleotide substitutions (including reversal of a chain-terminating codon), and a tenfold tandem reiteration of the coding sequence (Nurminsky et al. 1998). The origin of this new gene was a unique event, and many of the mutations involved in its subsequent evolution were unique. Such events are presently beyond the scope of population genetics, because it has not been possible to formulate general principles governing events that occur only once. The principles of population genetics apply primarily to the patterns and processes that result from recurrent events, and so it is to recurrent mutation that we now turn.

Forward Mutation

A useful model for analyzing changes in allele frequency due to recurrent mutation is the Hardy-Weinberg model with the assumption of no mutation relaxed. Consider, therefore, a gene with two alleles A and a, and suppose that the mutation rate per generation from A to a is μ. This process is called **forward mutation** if A is the prevalent wildtype allele. A mutation rate of μ per generation means that, in the transition from any one generation to the next, a fraction μ of A alleles undergo mutation and become a alleles, whereas a fraction $1 - \mu$ of A alleles escape mutation and remain A. Therefore, if p is the allele frequency of A in any generation, the allele frequency in the subsequent generation will be $p' = p(1 - \mu)$. Hence

$$p_t = p_{t-1}(1-\mu) = p_{t-2}(1-\mu)^2 = \cdots = p_0(1-\mu)^t \qquad (2.1)$$

where t is time in generations and p_0 is the allele frequency in the original population.

Suppose also that $p_0 \approx 1$ (the initial population is nearly fixed for A) and that t is not too large relative to $1/\mu$. Then $p_t \approx p_0 - t\mu$, and so the allele frequency q_t of the mutant a allele is given to a good approximation by

$$q_t = q_0 + t\mu \qquad (2.2)$$

This equation implies that the frequency of the mutant a allele increases linearly with time and that the slope of the line equals μ. Because μ is small, the linear increase in q is difficult to detect experimentally except in very large

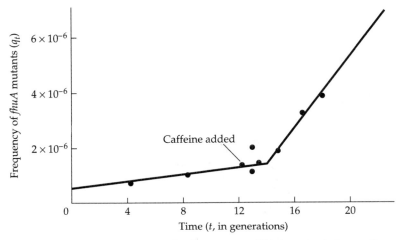

Figure 2.1   Linear increase in the frequency of *fhuA* mutants in a culture of *Escherichia coli* due to recurrent mutation. The mutants are detected via resistant to the bacteriophage T5. The mutation rate is estimated as the slope, in this case 7.2×10^{-8} per generation initially, and 6.6×10^{-7} per generation after the addition of caffeine. (From Novick 1955.)

populations of the magnitude achievable in experimental populations of bacteria. An example is shown in Figure 2.1. Note the abrupt increase in mutation rate (indicated by the increase in slope) shortly after the addition of caffeine, a bacterial mutagen.

Because spontaneous forward mutation rates are typically rather small (on the order of 10^{-4} to 10^{-6} mutations per allele per generation), the tendency for allele frequencies to change as a result of recurrent mutation (**mutation pressure**) is very small over the course of a few generations. On the other hand, the cumulative effects of mutation over long periods of time can become appreciable, as shown in Figure 2.2, where the solid curve has been calculated from Equation 2.1 for $\mu = 10^{-5}$. Because A undergoes recurrent mutation to a, but a never reverse mutates to A, the allele frequency of A eventually goes to zero, but very slowly. For realistic values of μ, it requires $t = 0.693/\mu$ generations to decrease the value of p by half, which in Figure 2.2 means a half-life of 69,300 generations.

Reversible Mutation

Something else happens when mutation is reversible. The population eventually reaches a stable **equilibrium** at which both A and a are present at frequencies that remain constant through time. Let us examine the same model as in the previous section, but allow reverse mutation from a to A at the rate

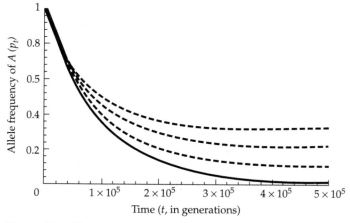

Figure 2.2 Changes in allele frequency under irreversible (solid curve) or reversible (dashed curves) mutation.

v per generation. In any generation, an A allele could have only two origins relative to the previous generation. It could have been an A allele in the previous generation that escaped mutation to a, or it could have been an a allele in the previous generation that underwent reverse mutation to A. These possibilities are captured by the equation

$$p_t = p_{t-1}(1 - \mu) + q_{t-1}v \tag{2.3}$$

A little algebraic manipulation puts this in the form

$$p_t - \frac{v}{\mu+v} = \left(p_{t-1} - \frac{v}{\mu+v} \right)(1-\mu-v) \tag{2.4}$$

which, by repeated substitution as in Equation 2.1, leads to

$$p_t - \frac{v}{\mu+v} = \left(p_0 - \frac{v}{\mu+v} \right)(1-\mu-v)^t \tag{2.5}$$

What happens to the allele frequencies in the long run is shown by the dashed lines in Figure 2.2. The frequency of A no longer goes to zero but to an equilibrium value given by

$$\hat{p} = \frac{v}{\mu+v} \tag{2.6}$$

In the dashed curves, $v = 0.1\mu$, 0.3μ, and 0.5μ, corresponding to the equilibrium frequencies $p = 1/11$, $3/13$, and $5/15$, respectively. The equilibrium is called a **stable equilibrium** because the allele frequencies converge to the equilibrium given in Equation 2.6 irrespective of the starting frequencies. Once again, for realistic values of the mutation rates, it takes a long time to reach the vicinity of the equilibrium.

Remote Inbreeding in a Finite Population

The mutation models considered so far assume an infinite population size, but actual populations are limited in size to some finite number of individuals. To be able to deal with finite populations we will extend the concept of allele identity by descent (IBD) introduced in Chapter 1. We previously defined IBD as the probability that the two alleles at a locus in an individual are identical in DNA sequence by virtue of being derived by DNA replication from a single ancestral allele. This definition is a key concept in the theory of inbreeding. Now we wish to generalize IBD to include *any* two alleles chosen at random from a population. The alleles need not necessarily be present in the same individual.

One important implication of a finite population persisting through many generations is that, eventually, every member of the population becomes related in some degree or another to every other member of the population. Everyone is related because, the population being limited in size, any two individuals must share at least one recent or remote common ancestor. It follows that mating pairs must be related in some degree or another, even if the mating pairs are formed "at random." This constitutes a type of inbreeding due to the sharing of remote ancestors, which we shall call **remote inbreeding** to distinguish it from the close inbreeding between immediate relatives discussed in Chapter 1.

Remote inbreeding does not cause a departure from HWE because of increased IBD of the alleles within each individual because, with remote inbreeding, mating pairs are still formed at random within the pool of potential partners available in the finite population. The main consequence of remote inbreeding is that, although HWE is maintained in the finite population through all generations, the probability of IBD of two randomly chosen alleles in the population steadily increases.

The reason for the increase in IBD is shown in Figure 2.3. The alleles are labeled α_1, α_2, α_3, and so forth in order to mask their identities as either A or a, because here we are interested only in IBD. We denote the probability that any two randomly chosen alleles are IBD in generation t as F_t. In much of the population genetics literature the probability of IBD due to remote inbreeding is called the **fixation index** and symbolized as F_{ST} or G_{ST} to distinguish it from the conventional inbreeding coefficient due to close inbreeding. The ST is omitted in Figure 2.3 to avoid a proliferation of subscripts.

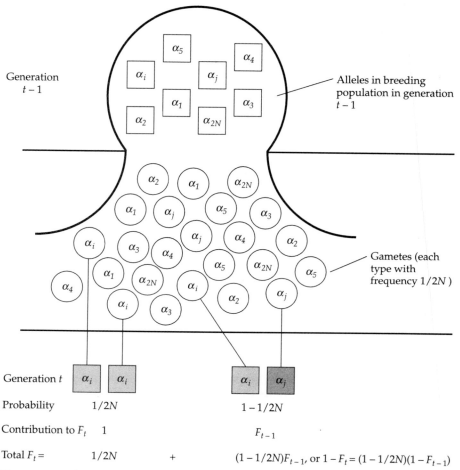

Generation
$t-1$

Alleles in breeding
population in generation
$t-1$

Gametes (each
type with
frequency $1/2N$)

Generation t	α_i	α_i		α_i	α_j
Probability		$1/2N$			$1-1/2N$
Contribution to F_t		1			F_{t-1}
Total $F_t =$		$1/2N$	$+$		$(1-1/2N)F_{t-1}$, or $1-F_t=(1-1/2N)(1-F_{t-1})$

Figure 2.3 Identity by descent of alleles due to random sampling in a finite population.

In Figure 2.3, we assume that a population of size N diploid individuals in generation $t-1$ generates an infinite pool of gametes, from which $2N$ are chosen to be represented in the next generation. Consider any two randomly chosen gametes, which may not necessarily be present in the same zygote. There are only two possibilities for the chosen pair of alleles with regard to their identity in the previous generation. They could either derive from exactly the same allele (α_i) in breeding adults of the previous generation; this event has a probability of $1/(2N)$. Or they could derive from two different alleles (α_i and α_j) in breeding adults of the previous generation; this outcome has a probability of $1 - 1/(2N)$. In the first case $(\alpha_i \alpha_i)$, the chosen pair of alleles is certainly IBD. In the second case $(\alpha_i \alpha_j)$, they are IBD in the present generation only if they were IBD in the previous generation, and this has the probability F_{t-1}. Putting these possibilities together, the relation between F_t and F_{t-1} is

$$F_t = \frac{1}{2N} + \left(1 - \frac{1}{2N}\right)F_{t-1} \tag{2.7}$$

To both sides first multiply by -1 and then add $+1$. This yields

$$1 - F_t = \left(1 - F_{t-1}\right)\left(1 - \frac{1}{2N}\right) \tag{2.8}$$

and again by the method of successive substitutions we obtain

$$1 - F_t = \left(1 - F_0\right)\left(1 - \frac{1}{2N}\right)^t \tag{2.9}$$

Equation 2.9 shows that F_t goes to 1 (complete IBD) at a rate that depends on the reciprocal of the population size. Eventually, all of the alleles in the finite population become IBD. How can this happen? It happens because the allele frequencies of A and a change randomly from generation to generation, and eventually either a single A allele present in the original population, or a single a allele present in the original population, becomes fixed.

Equilibrium Heterozygosity with Mutation

The sampling process in Figure 2.3 leads ultimately to total homozygosity only in the absence of mutation. When mutation occurs, the population can regain some heterozygosity in each generation because alleles lost due to random sampling can be replaced by different alleles generated by new mutations. Whenever an allele changes state by mutation, it breaks the chain of IBD between itself and its ancestors, so F_t no longer necessarily goes to 1. In order to keep track of the new mutations, we will suppose that every time a new mutation occurs it creates a novel allele that does not already exist in the population. This is called the **infinite-alleles model** of mutation. The infinite-alleles model is but one way to specify the characteristics of new mutations. Although it represents a somewhat simplified view of mutation, it nevertheless provides a useful standard of comparison for other models or for observed allele frequencies.

The calculation of F_t in Figure 2.3 is still correct for this model, provided that neither allele (α_i or α_j) undergoes mutation in the passage of one generation. Therefore, under the infinite-alleles model,

$$F_t = \left[\frac{1}{2N} + \left(1 - \frac{1}{2N}\right) F_{t-1} \right] (1 - \mu)^2 \tag{2.10}$$

In this case, F_t does not go to 1 but to an equilibrium value $\hat{F}$ found by solving Equation 2.9 with $F_t = F_{t-1} = \hat{F}$. The result sought is, to an excellent approximation,

$$\hat{F} = \frac{1}{4N\mu + 1} \tag{2.11}$$

Furthermore, because each new mutation creates a unique allele, $\hat{F}$ is also the equilibrium frequency of homozygous genotypes. Hence the equilibrium frequency of heterozygous genotypes, $\hat{H}$, equals $1 - \hat{F}$, or

$$\hat{H} = 1 - \hat{F} = \frac{4N\mu}{4N\mu + 1} = \frac{\theta}{1 + \theta} \tag{2.12}$$

where $\theta = 4N\mu$ is a convention often used in population genetics theory because N and μ cannot usually be estimated separately.

To apply Equation 2.12, consider that allozyme polymorphisms in *Drosophila* have a mean heterozygosity of about 0.14; hence $\langle \theta \rangle = 0.163$ for allozyme alleles in *Drosophila*, but the 95% confidence interval around this estimate is very large (see Figure 1.8). In most higher organisms, allozyme heterozygosities are in the range 6% to 19%, yielding $\langle \theta \rangle$ values in the range 0.06–0.23 (Lewontin 1974). Taking $\theta = 0.163$ and $N = 10^6$ for *Drosophila* (Akashi 1997) results in $\mu = 4 \times 10^{-8}$ for allozyme alleles. This is undoubtedly lower than the actual mutation rate to new allozyme alleles and suggests that the great majority of electrophoretic enzyme variants must be harmful and eliminated so rapidly by selection that they do not persist for long as allozyme polymorphisms.

MIGRATION

The term **migration** refers to the movement of individuals among subpopulations. It is a sort of genetic glue that holds subpopulations together and sets a limit to how much genetic divergence can occur. To understand the homogenizing effects of migration, it is useful to study migration in simple models of population structure.

The Island Model of Migration

In the **island model** of migration, a large population is split into many subpopulations dispersed geographically like islands in an archipelago. Examples of island population structure might include fish in freshwater lakes or slugs in dispersed garden plots. Each subpopulation is assumed to be so large that random changes in allele frequency can be neglected. Consider two alleles A and a with allele frequencies that differ among a set of subpopulations. Migration is assumed to occur in such a way that the allele frequencies among migrants equal the average values among subpopulations, designated $\bar{p}$ and $\bar{q}$. The amount of migration is measured by the parameter m, which equals the probability that a randomly chosen allele in any subpopulation comes from a migrant. For any randomly chosen allele in any subpopulation in generation t, the allele could have come from the same subpopulation in generation $t-1$ (with probability $1-m$), in which case it is an A allele with probability p_{t-1}. Alternatively, the allele could have come from a migrant in generation $t-1$ (with probability m), in which case it is an A allele with probability $\bar{p}$. Since all assumptions of the HWE are in force except that of no migration, $\bar{p}$ stays the same in all generations.

Altogether,

$$p_t = p_{t-1}(1-m) + \bar{p}m \tag{2.13}$$

Equation 2.13 is similar in form to Equation 2.3 for reversible mutation. Its solution in terms of p_0 is

$$p_t = \bar{p} + (p_0 - \bar{p})(1-m)^t \tag{2.14}$$

where p_0 is the initial frequency of A in the subpopulation of interest. Suppose for example that there were only two subpopulations, with initial allele frequencies of A of 0.2 and 0.8 and with $m = 0.10$. This value of m means that 10% of the individuals in either subpopulation in any generation are migrants in which the allele frequency of A equals $\bar{p} = (0.2 + 0.8)/2 = 0.5$. Suppose we wish to deduce the allele frequency of A in the two populations after 10 generations. For the population with initial allele frequency 0.2, we substitute $p_0 = 0.2$, $\bar{p} = 0.5$, and $m = 0.10$ into Equation 2.14 to obtain $p_{10} = 0.5 + (0.2 - 0.5)(1 - 0.10)^{10}$, or $p_{10} = 0.395$. The allele frequency in the other population is obtained similarly as $p_{10} = 0.5 + (0.8 - 0.5)(1 - 0.10)^{10} = 0.605$.

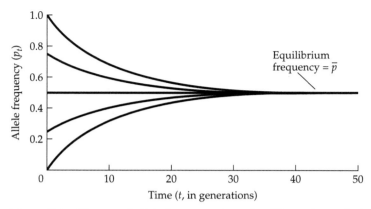

Figure 2.4   Change of allele frequency in each of five subpopulations in the island model of migration.

A graphical example using Equation 2.18 is shown in Figure 2.4, where again $m = 0.10$, but this time with five subpopulations having initial frequencies 1, 0.75, 0.50, 0.25, and 0. Note how rapidly the allele frequencies converge to an equilibrium value equal to the average among the initial subpopulations, in this case 0.5.

How Migration Limits Genetic Divergence

In this section we again invoke the extended concept of allele identity by descent (IBD) to examine the offsetting effects of migration and random change in allele frequency due to finite population size. Recall that the extended definition defines F_t (the fixation index) as the probability of IBD between any two alleles chosen at random from the same subpopulation. Among a set of subpopulations that do not exchange migrants, F_t gradually increases in each subpopulation. This means that the allele frequencies change among the subpopulations—in other words the subpopulations undergo genetic divergence. In the absence of migration, the genetic divergence continues until ultimately each subpopulation becomes fixed for either the A allele or the a allele due to the accumulated effects of remote inbreeding.

It is remarkable how little migration is required to prevent significant genetic divergence due to remote inbreeding within the subpopulations. The effect can be seen quantitatively by considering the model in Figure 2.3, but permitting migration at a rate m according to the island model. The expression for F_t derived in Figure 2.3 is still valid, provided that neither allele α_i or allele α_j has been replaced by a migrant allele. Therefore

$$F_t = \left[\frac{1}{2N} + \left(1 - \frac{1}{2N}\right)F_{t-1}\right](1-m)^2 \tag{2.15}$$

Equation 2.15 is identical to Equation 2.10 except that the parameter μ is replaced with the parameter m. Hence, for reasonably small values of m, the equilibrium fixation index is approximately

$$\hat{F} = \hat{F}_{ST} = \frac{1}{4Nm+1} \tag{2.16}$$

Not surprisingly, this looks exactly like Equation 2.11 except that Nm replaces $N\mu$. This emphasizes the theoretical similarity between the effects of mutation and migration. The practical difference is that rates of migration between subpopulations are typically very much greater than rates of mutation, so the implications of Equations 2.11 and 2.16 are very different.

Equation 2.16 implies that $\hat{F}$ decreases as the number of migrants increases, but the decrease is extremely rapid, as shown in Figure 2.5. The dashed horizontal lines demarcate regions advised by Wright (1978) for qualitative interpretation of the fixation index in terms of degree of genetic divergence among the subpopulations. A value of Nm as small as 2 already brings $\hat{F}$ into the zone of "moderate divergence." Because N is the size of each subpopulation, and m

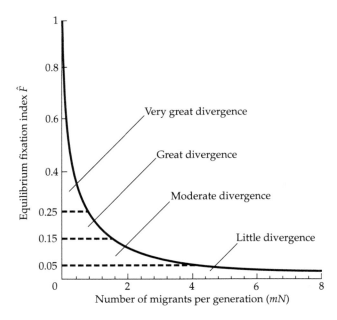

Figure 2.5 Equilibrium fixation index ($\hat{F}$ or $\hat{F}_{ST}$) against the number of migrants per generation, assuming the island model of migration.

Table 2.1 Estimates of Nm and $\hat{F}_{ST}$

Species	Type of organism	Estimated Nm	Estimated $\hat{F}_{ST}$
Stephanomeria exigua	Annual plant	1.4	0.152
Mytilus edulis	Mollusc	42.0	0.006
Drosophila willistoni	Insect	9.9	0.025
Drosophila pseudoobscura	Insect	1.0	0.200
Chanos chanos	Fish	4.2	0.056
Hyla regilla	Frog	1.4	0.152
Plethodon ouachitae	Salamander	2.1	0.106
Plethodon cinereus	Salamander	0.22	0.532
Plethodon dorsalis	Salamander	0.10	0.714
Batrachoseps pacifica ssp. 1	Salamander	0.64	0.281
Batrachoseps pacifica ssp. 2	Salamander	0.20	0.556
Batrachoseps campi	Salamander	0.16	0.610
Lacerta melisellensis	Lizard	1.9	0.116
Peromyscus californicus	Mouse	2.2	0.102
Peromyscus polionotus	Mouse	0.31	0.446
Thomomys bottae	Gopher	0.86	0.225

Source: Data from Slatkin 1985.

is the probability that an individual in a subpopulation is a migrant, Nm corresponds to the absolute *number* of migrants into each subpopulation each generation. This means that, independent of subpopulation size, two or more migrants per generation severely restricts genetic divergence.

How large is Nm in natural populations? One method of estimating genetic migration relies on the finding that, in theoretical models, the logarithm of Nm decreases approximately as a linear function of the average frequency of "private" alleles that are unique to individual samples from the subpopulations (Slatkin 1985). Estimates based on this method and the resulting equilibrium values of F_{ST} are summarized in Table 2.1. There is obviously considerable variation among organisms, but many of the values of Nm are approximately 2 or smaller, which gives considerable scope for genetic divergence resulting from random genetic drift.

Wahlund's Principle

We noted in Chapter 1 that admixture of two or more HWE populations with differing allele frequencies produces a mixed population that has a deficiency of heterozygous genotypes relative to the frequency expected with HWE for the average allele frequencies. This phenomenon is illustrated

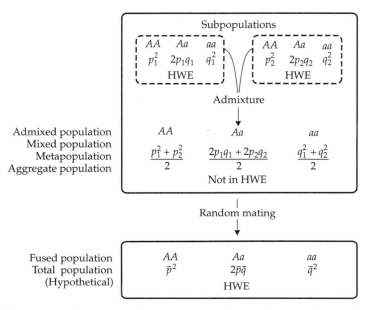

Figure 2.6 A metapopulation consists of a set of more or less isolated subpopulations. Alternative terms are admixed population, mixed population, or aggregate population. This diagram illustrates that the genotype frequencies among each of a group of subpopulations can be in HWE, but the genotype frequencies in the metapopulation are not in HWE. If the individuals in the subpopulations are combined into one large, random-mating population, the result is a fused population often called the total population.

quantitatively for two subpopulations in Figure 2.6. Each subpopulation itself is in HWE for its own allele frequencies, but the admixed population (metapopulation) composed of the aggregate of subpopulations is not in HWE. Relative to an HWE population with allele frequencies equal to the average of those among the subpopulations, the aggregate population contains too few heterozygous genotypes. The flip side of the coin is that the metapopulation contains too many homozygous genotypes.

These effects are conveniently measured in terms of the variance in allele frequency among the subpopulations. The **variance** in any quantity, usually symbolized σ^2, is defined as the average of the squared deviation of each value from the overall mean, or, in terms of the recessive allele frequencies among the subpopulations in Figure 2.6,

$$\sigma^2 = \frac{\left(q_1 - \bar{q}\right)^2 + \left(q_2 - \bar{q}\right)^2}{2} \tag{2.17}$$

where $\bar{q} = (q_1 + q_2)/2$ is the mean. When there are only two alleles, $q = 1 - p$, from which it easily follows that the variance of p equals the variance of q. Completing the squares in Equation 2.17 and simplifying leads to an alternative expression for the variance

$$\sigma^2 = \overline{q^2} - \bar{q}^2 \tag{2.18}$$

where the symbol $\overline{q^2}$ means the mean of the squares, $\overline{q^2} = (q_1^2 + q_2^2)/2$.

Using Equation 2.18, we can compare the genotype frequencies in the metapopulation (equal to the average genotype frequencies among the subpopulations) with the genotype frequencies that would be expected in a hypothetical total population in HWE with the average allele frequencies. Define R_S as the frequency of homozygous recessive genotypes in the metapopulation and R_T as the frequency that would be expected in the totally fused population in HWE. From Figure 2.6 we have

$$R_S - R_T = \overline{q^2} - \bar{q}^2 = \sigma^2 \tag{2.19}$$

This equation is **Wahlund's principle.** It states that the average frequency of homozygous recessive genotypes among a group of subpopulations is always greater than the frequency of homozygous recessive genotypes that would be expected with random mating, and the excess is numerically equal to the variance in the recessive allele frequency. To take a specific example, imagine a population of gray squirrels that by chance has acquired a frequency of recessive albinism equal to 16%. In a nearby forest is another population in which the albino allele is absent, so the allele frequency in this population is 0. Overall, the average frequency of albinos in the two populations is $(0.16 + 0)/2 = 8\%$. Were the two populations to fuse and undergo random mating, the allele frequency of the albino allele in the fused population would be $\left[\sqrt{(0.16)} + \sqrt{(0)}\right]/2 = 0.2$, and the frequency of the homozygous recessive genotype would equal $(0.2)^2 = 4\%$, which is in fact less than the average of the separate subpopulations. Furthermore, the variance in allele frequency equals $[(0.4 - 0.2)^2 + (0 - 0.2)^2]/2 = 0.04$, which does equal the reduction in frequency of the homozygous recessive.

Hierarchical Population Structure

An equation similar to 2.19 also applies to the frequency D of homozygous dominant genotypes, namely,

$$D_S - D_T = \overline{p^2} - \bar{p}^2 = \sigma^2 \tag{2.20}$$

and since an excess of homozygous genotypes must be matched exactly by a deficiency of heterozygous genotypes, it follows that

$$H_S - H_T = -2\sigma^2 \qquad (2.21)$$

But recall from Chapter 1 that we can define the inbreeding coefficient as the deficiency of heterozygous genotypes in an inbred population, relative to a population in HWE. In this case the "inbreeding coefficient" is the fixation index F_{ST}, and the "inbred population" is the metapopulation; nevertheless,

$$F_{ST} = \frac{H_T - H_S}{H_T} \qquad (2.22)$$

Substituting from Equation 2.21 and noting that $H_T = 2\bar{p}\bar{q}$ leads to

$$F_{ST} = \frac{\sigma^2}{\bar{p}\bar{q}} \qquad (2.23)$$

Equation 2.23 is not just a trick of algebraic manipulation. It is a fundamental relation in population genetics that connects the fixation index in a metapopulation with the variance in allele frequencies among the subpopulations. The fixation index can also be interpreted in terms of the probability of IBD, and therefore genotype frequencies in the metapopulation can be derived using the same arguments as used for close inbreeding in Equations 1.16–1.19. The results are the genotype frequencies

$$
\begin{array}{lll}
AA & \bar{p}^2 + \bar{p}\bar{q}F_{ST} & = \bar{p}^2(1 - F_{ST}) + \bar{p}F_{ST} \\
Aa & 2\bar{p}\bar{q} - 2\bar{p}\bar{q}F_{ST} & = 2\bar{p}\bar{q}(1 - F_{ST}) \\
aa & \bar{q}^2 + \bar{p}\bar{q}F_{ST} & = \bar{q}^2(1 - F_{ST}) + \bar{q}F_{ST}
\end{array}
\qquad (2.24)
$$

This result has always struck me as one of the deep paradoxes of population genetics. It says that there is inbreeding in the metapopulation composed of the aggregate of subpopulations, even though each subpopulation itself is undergoing random mating and is in HWE. The reason for the paradox is that the population as a whole is *not* undergoing random mating. There is remote inbreeding because matings occur only within subpopulations, and because each subpopulation is finite in size, the level of inbreeding as measured by F_{ST} gradually builds up. This kind of population structure is called a **hierarchical population structure** because it is composed in a hierarchy of subpopulations within larger aggregates.

NATURAL SELECTION

As Darwin correctly proposed over a century ago, natural selection is the driving force of evolution. By **selection** we mean inherited differences in the ability of organisms to survive and reproduce, so that through time the genotypes that are superior in survival and reproduction increase in frequency in the population. Selection is the principal process that results in greater adaptation of organisms to their environment, because what we really mean by **adaptation** is the acquisition of traits that enhance survival and reproduction in a given environment. In any species, genetic variation produced by mutation is organized, maintained, eliminated, or dispersed among subpopulations according to a complex balance of migration, random changes in allele frequency, and selection acting on many traits.

Haploid Selection

Selection in its simplest form occurs in haploid organisms such as bacteria or yeast. The main concepts in modeling haploid selection are summarized in a numerical example in Figure 2.7. We assume two competing genotypes A and a with relative frequencies p and q at some specified time. Each little box represents an organism, and the alleles are represented by open and filled circles. To simplify matters we assume nonoverlapping generations, so that a population of cells undergoes reproduction to produce haploid progeny of the next generation and then immediately dies. Haploid organisms like yeast may undergo mating to produce ephemeral diploids that undergo meiosis to produce the next generation of haploid progeny. Asexual organisms reproduce without mating. In either case, the haploid progeny undergo a process of maturation in which some of the organisms fail to survive. In the example we assume that the death rates are 1/10 for A and 4/10 for a, so the probability of survival, which is called the **viability** or **survivorship,** depends on the genotype.

The overall ability of an organism to survive and reproduce constitutes its **fitness.** In most models in population genetics, for the sake of simplicity the only component of fitness that is considered is viability. Accordingly, we estimate the fitnesses of the A and a genotypes as 9/10 and 6/10. These are the **absolute fitnesses** because they measure the survivorship of each genotype separately. In the real world, it is virtually impossible to estimate absolute fitness because it requires knowing the absolute number of each genotype present immediately after reproduction. However, it *is* possible to estimate **relative fitness,** which means the ability of one genotype to survive rel-

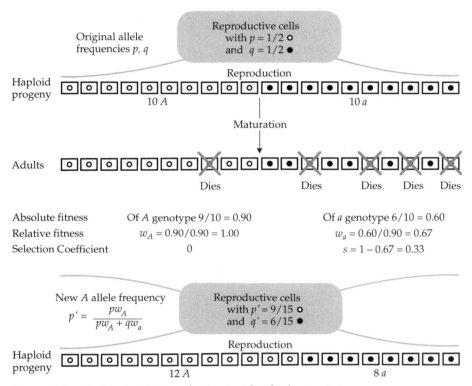

Figure 2.7 Model of viability selection in a haploid organism.

ative to another genotype taken as a standard. In the example in Figure 2.7 we take A as the standard with relative fitness $w_A = 1$, yielding a relative fitness for a of $w_a = 0.67$, which equals the ratio of the viabilities of the genotypes. This means that, for every A genotype that survives, 0.67 a genotypes survive (6 a versus 9 A in the numerical example). An equivalent way to measure the relative fitness of a is according to the difference $1 - (w_a/w_A)$, which in this case is $1 - 0.67 = 0.33$; this parameter is called the **selection coefficient** against a, which is conventionally denoted s.

Figure 2.7 spells out the details of how selection changes the allele frequencies from one generation to the next. In the example, the first generation of haploid progeny included 10 A alleles and 10 a alleles prior to selection, but after selection only 9 A and 6 a alleles remain. These frequencies are assumed to be maintained in the haploid progeny of the next generation, which means that differences in fertility between the genotypes are negligible. Therefore, in the next generation the genotype frequencies before selection are $p' = 9/15$ and $q' = 6/15$. The more general equation for allele frequency change under haploid selection is

$$p' = \frac{pw_A}{pw_A + qw_a} = \frac{p}{p + q(1-s)} = \frac{p}{1-qs} \tag{2.25}$$

which should be checked by direct substitution of $p = 1/2$, $q = 1/2$, $w_A = 1$, $w_a = 2/3$, and $s = 1/3$. This is one of the few selection equations in population genetics that has an analytical solution, namely,

$$p_t = \frac{p_0}{p_0 + q_0(1-s)^t} \tag{2.26}$$

To obtain the solution, note from Equation 2.25 that $q' = q(1-s)/(1-qs)$ because $p' + q' = 1$. Hence, by the method of successive substitutions,

$$\frac{p_t}{q_t} = \left(\frac{p_{t-1}}{q_{t-1}}\right)\left(\frac{1}{1-s}\right) = \cdots = \left(\frac{p_0}{q_0}\right)\left[\frac{1}{(1-s)^t}\right] \tag{2.27}$$

from which Equation 2.26 follows immediately. However, taking the natural logarithm of both sides of Equation 2.27 tells us something else:

$$\ln\left(\frac{p_t}{q_t}\right) = \ln\left(\frac{p_0}{q_0}\right) - t\ln(1-s) \cong \ln\left(\frac{p_0}{q_0}\right) + st \tag{2.28}$$

where the approximation $\ln(1-s) \approx -s$ is quite good even for s as large as 20%. Equation 2.28 means that, if s is not too large, a plot of $\ln(p/q)$ against time in generations, t, should be linear with a slope equal to the value of s. This is one approach by which the selection coefficient can be estimated. Furthermore, linearity of the plot implies that the relative fitnesses are constant through time.

A real example of such a plot is shown in Figure 2.8, which shows the result of competition between pairs of genotypes of E. coli that differ in a single electrophoretic enzyme variant but are otherwise genetically identical (Hartl and Dykhuizen 1981). The open dots are the data from a pair of strains differing in fitness by a selection coefficient $s = 3.23\%$, with selection favoring the strain designated A. The solid dots are data for a pair of strains differing in fitness by 6.14%, with selection favoring the strain designated a.

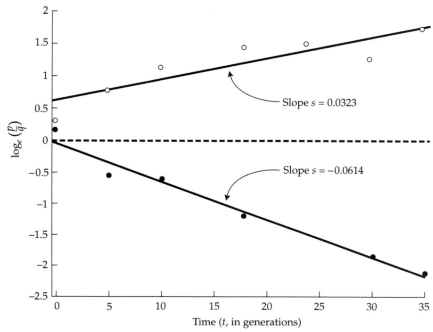

Figure 2.8   Selection between strains of *Escherichia coli* that are genetically identical except for alleles encoding different allozymes of 6-phosphogluconate dehydrogenase. The selection coefficient is estimated as the slope of the straight line. The allele designations are arbitrary. In one pair of strains (open circles), the allele designated *A* is favored; in the other pair (filled circles), the allele designated *a* is favored. (Data from Hartl and Dykhuizen 1981.)

Diploid Selection

The concepts of diploid selection are identical to those for haploid selection, but there are complications resulting from Mendelian segregation and heterozygous genotypes. These are illustrated in Figure 2.9, where again each little box represents an organism but each organism contains two alleles of the gene. Selection is assumed to occur on the diploid genotypes, not on the gametes, and so with random mating a large pool of gametes with allele frequencies p and q produces a pool of zygote genotypes in HWE for the same allele frequencies. Viability selection occurs between fertilization and maturation. There are three absolute fitnesses corresponding to the genotypes AA, Aa, and aa, but taking AA as the standard there are just two relative fitnesses: w_{Aa} is the fitness of Aa relative to AA, and w_{aa} is the fitness of aa relative AA. In this example, the relative fitnesses with $w_{AA} = 1$ as the standard are

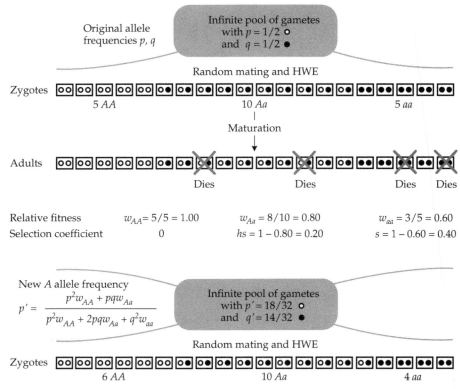

Figure 2.9 Model of viability selection in a diploid organism.

$w_{Aa} = 8/10$ and $w_{aa} = 6/10$. By analogy with haploid selection, we can also define a selection coefficient against Aa and against aa. The selection coefficient against homozygous aa is denoted s, and $s = 1 - (w_{aa}/w_{AA})$, which in this case equals 0.40. The selection coefficient against the heterozygous genotype is usually expressed as a multiple h of s, and so $hs = 1 - (w_{Aa}/w_{AA})$, which in this case is 0.20. Hence $h = hs/s = 0.20/0.40 = 0.50$. The parameter h is called the **degree of dominance** of a, and it is a convenient parameter in diploid models because:

- $h = 0$ means that a is recessive to A.
- $h = 1/2$ means that the heterozygous fitness is the arithmetic average of the homozygous fitnesses; in this case, the effects of the alleles are said to be **additive effects.** Additive effects are exemplified in Figure 2.9. This case is also called **genic selection.**
- $h = 1$ means that a is dominant to A.

As we shall see below, it is also possible the $h < 0$ or $h > 1$, but for the moment we will consider only the range $0 \leq h \leq 1$.

In Figure 2.9, the genotype frequencies after the viability selection are 5 *AA*, 8 *Aa*, and 3 *aa*, which corresponds to allele frequencies of 18 *A* and 14 *a*. As before, we assume that these remain unchanged in the gametes and zygotes of the next generation (no differential mating success or fertility selection). It may be verified by numerical substitution that the allele frequency in the next generation prior to selection is given by

$$p' = \frac{p^2 w_{AA} + pq w_{Aa}}{p^2 w_{AA} + 2pq w_{Aa} + q^2 w_{aa}} \tag{2.29}$$

where the $pq w_{Aa}$ in the numerator is not multiplied by 2 because only half the gametes from an *Aa* genotype carry *A*. Note that the denominator of Equation 2.29 is the average fitness in the population, usually symbolized $\bar{w}$

$$\bar{w} = p^2 w_{AA} + 2pq w_{Aa} + q^2 w_{aa} \tag{2.30}$$

Equation 2.29 does not usually have an analytical solution, and for this reason it is usually more useful to calculate the difference in allele frequency between two successive generations $p' - p$, which is usually symbolized Δp. This can be found by subtracting p from both sides of Equation 2.29, multiplying p on the right-hand side by $\bar{w}/\bar{w}$ to obtain a common denominator, and simplifying. The result is

$$\Delta p = \frac{pq[p(w_{AA} - w_{Aa}) + q(w_{Aa} - w_{aa})]}{\bar{w}} \tag{2.31}$$

To illustrate the use of these equations we will use data on change in the frequency of the allele *Cy* (*Curly wings*) in a laboratory population of *D. melanogaster* (Teissier 1942). These are plotted in Figure 2.10 as the frequency of *Cy/+* adults in the population. Since *Cy/Cy* homozygotes die as embryos, the allele frequency of *Cy* equals the frequency of *Cy/+* adults divided by 2. In order to be able to apply the equations directly, we set *A* to correspond to the *Cy* allele and *a* to the + allele. In this case *aa* is the standard genotype for calculating the relative fitnesses, so $w_{aa} = 1$. Because *AA* (*Cy/Cy*) is lethal, $w_{AA} = 0$. The best-fitting curve for these data is Equation 2.29 with $w_{Aa} = 0.50$ (Wright 1977), thus the effects of *Cy* on fitness are additive. The expected genotype frequencies can be calculated recursively from Equation 2.29. The

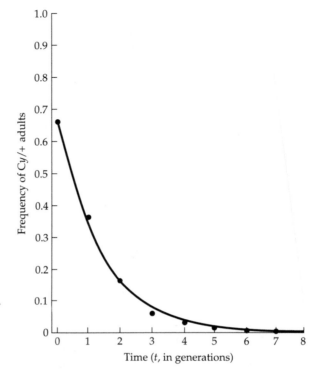

Figure 2.10 Change in frequency of adult *Drosophila melanogaster* heterozygous for the dominant mutation *Cy* (*Curly wings*) in an experimental population. (Data from Teissier 1942.)

initial population was established with a frequency of *Cy/+* adults of 2/3, so $p = 1/3$ and $q = 2/3$. In the second generation we expect

$$p' = \frac{(1/3)^2(0) + (1/3)(2/3)(0.50)}{(1/3)^2(0) + 2(1/3)(2/3)(0.50) + (2/3)^2(1.00)} = \frac{1}{6}$$

This predicts the frequency of *Cy/+* adults in the next generation as $2p' = 1/3$ = 0.33, which is reasonably close to the observed value of 0.368.

Time Required for Changes in Gene Frequency

In Equation 2.31, if selection is weak enough that $\bar{w} \approx 1$, then the expression for Δp can be used to derive approximate equations for the time required to effect any specified change in allele frequency. These approximations require integral calculus (the idea is to treat Δp as the derivative (dp/dt) and then integrate), but for our purposes only the final answers are of interest. The answers are most easily presented in terms of the selection coefficient and the degree of dominance. With these as the parameters and $\bar{w} \approx 1$, Equation 2.31 becomes

$$\Delta p = pqs\left[ph + q(1-h)\right] \tag{2.32}$$

The time t required for the allele frequency of A to change from p_0 to p_t in each of three cases of special interest is the following.

1. *A* **is a favored dominant.** In this case $h = 0$ and $\Delta p = pq^2 s$, which yields

$$\ln\left(\frac{p_t}{q_t}\right) + \left(\frac{1}{q_t}\right) = \left[\ln\left(\frac{p_0}{q_0}\right) + \frac{1}{q_0}\right] + st \tag{2.33}$$

2. *A* **is favored and the alleles are additive.** Here $h = 1/2$ and $\Delta p = pqs/2$. The integral is

$$\ln\left(\frac{p_t}{q_t}\right) = \ln\left(\frac{p_0}{q_0}\right) + \left(\frac{s}{2}\right)t \tag{2.34}$$

The equation for additive alleles is identical to Equation 2.28 for haploid selection if, in the haploid model, we let $w_A = 1$ and $w_a = 1 - (s/2)$, and stipulate that s is small.

3. *A* **is a favored recessive.** In this case $h = 1$ and $\Delta p = p^2 qs$, which produces

$$\ln\left(\frac{p_t}{q_t}\right) - \frac{1}{p_t} = \left[\ln\left(\frac{p_0}{q_0}\right) - \frac{1}{p_0}\right] + st \tag{2.35}$$

It is worthwhile to emphasize a particular implication of these equations for rare harmful recessives. This is the case when A is a favored dominant and $\Delta p = pq^2 s$. When a harmful allele is rare, q is close to 0, and q^2 is therefore extremely small. Consequently, an increase in s from a large value of, say, 0.5 to a still larger value of, say, 1, has a trivial effect on the change in allele frequency because, with q^2 so small, the actual value of s matters little. In other words, the change in allele frequency of a rare harmful recessive is slow whatever the value of the selection coefficient. For this reason, an increase in selection against rare homozygous recessive genotypes has almost no effect in changing the allele frequency. The implication for human population genetics is that the forced sterilization of rare homozygous recessive individuals—a procedure advocated in a number of naive eugenic programs to "improve" the "genetic quality" of human beings—is not only morally and ethically abhorrent, it is genetically unsound.

Changes in allele frequency from Equations 2.33–2.35 are shown in Figure 2.11. Note that the change in frequency of a favored dominant allele is slowest when the allele is common, and that the change in frequency of a favored recessive allele is slowest when the allele is rare. The explanation is the same in both cases. Rare recessive alleles are present mainly in heterozygous genotypes and thereby are hidden from selection. With a favorable additive allele, the initial increase in frequency is slower than that of a favored dominant, but the additive allele eventually overtakes and goes to fixation faster, because additive selection continues to distinguish between the homozygous and heterozygous genotypes.

Some of the most dramatic examples of evolution in action result from the natural selection for chemical pesticide resistance in natural populations of insects and other agricultural pests. In the 1940s, when chemical pesticides were first applied on a large scale, an estimated 7% of the agricultural crops in the United States were lost to insects. Initial successes in chemical pest management were followed by gradual loss of effectiveness. By 1985, more than 400 pest species had evolved significant resistance to one or more pesticides, and 13% of the agricultural crops in the United States were lost to insects. In many cases, significant pesticide resistance had evolved in 5–50 generations, irrespective of the insect species, geographical region, pesticide, frequency and method of application, and other seemingly important variables (May 1985). Equations 2.33–2.35 help to understand this apparent paradox, since many of the resistance phenotypes result from single mutant genes. The resistance genes are often partially or completely dominant, so Equations 2.33 and 2.34 are applicable. Prior to application of the pesticide, the allele frequency p_0 of the resistant mutant is usually close to 0. Application of the pesticide increases the allele frequency, sometimes by many

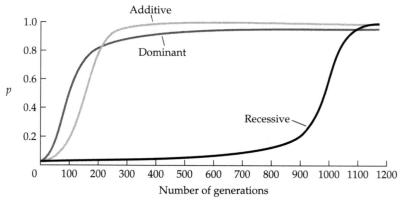

Figure 2.11 Change in frequency of a favored allele with dominant, additive, or recessive effects on fitness. In each case, there is a 5% difference in relative fitness between the homozygous genotypes.

orders of magnitude, but significant resistance is noticed in the pest population even before the allele frequency p_t increases above a few percent. Thus,
as rough approximations, we may assume that q_0 and q_t are both close
enough to 1 that $\ln(p_0/q_0) \approx \ln(p_0)$ and $\ln(p_t/q_t) \approx \ln(p_t)$. Using these approximations, Equation 2.34 (additive case) implies that $t \approx (2/s)\ln(p_t/p_0)$, and
Equation 2.33 (dominant case) implies that $t \approx (1/s)\ln(p_t/p_0)$. In many cases,
the ratio p_t/p_0 may range from 1×10^2 to perhaps 1×10^7, and s may typically
be 0.5 or greater. Over this wide range of parameter values, the time t is
effectively limited to 5–50 generations for the appearance of a significant
degree of pesticide resistance. It is instructive to calculate t for a few numerical examples. The time actually required will depend on such details as
population size and the extent of migration between local populations, and
the evolution of resistance due to multiple interacting genes may be expected to take somewhat longer than single-gene resistance.

Overdominance and Underdominance

An **equilibrium** value of p is a value for which $\Delta p = 0$; this means the allele
frequency remains at a constant value generation after generation. There are,
however, several types of equilibria, depending on what happens to allele
frequency when the allele frequency does not equal the equilibrium value.
Consider first the case when the initial allele frequency is near (but not equal
to) the equilibrium. If the allele frequency moves progressively farther away
from the equilibrium in subsequent generations, the equilibrium is said to be
unstable. If the allele frequency moves progressively closer to the equilibrium in subsequent generations, the equilibrium is said to be **locally stable.**
A locally stable equilibrium might also be **globally stable,** which means that,
whatever the initial allele frequency, it always moves progressively closer to
the equilibrium. In cases such as those exemplified by the HWE, in which
every allele frequency represents an equilibrium because, whatever its
value, the allele frequency does not change, the equilibria are said to be
semistable or **neutrally stable.**

These concepts of stability can be applied to the case of directional selection for the A allele governed by Equation 2.31. Directional selection for A
means that $w_{AA} \geq w_{Aa} \geq w_{aa}$ (but not both equalities). There are two equilibria, $p = 0$ and $p = 1$. If p is close to 0, p increases, so the equilibrium at $p = 0$ is
unstable. On the other hand, if p is near 1, it moves still closer to 1, so the
equilibrium at $p = 1$ is locally stable. Indeed, because p eventually goes to 1
whatever its initial value, the equilibrium at $p = 1$ is globally stable.

The various types of stability are important in discussing two further
cases that can occur when selection involves two alleles of a single gene.

A situation of **overdominance** or **heterozygote superiority** arises when
the heterozygous genotype has a greater fitness than either homozygous
genotype, or $w_{Aa} > w_{AA}$ and $w_{Aa} > w_{aa}$. With overdominance there is a third

equilibrium in addition to $p = 0$ and $p = 1$, because $p(w_{AA} - w_{Aa}) + q(w_{Aa} - w_{aa})$ can equal 0. The third equilibrium $\hat{p}$ can be found by solving $\hat{p}(w_{AA} - w_{Aa}) + \hat{q}(w_{Aa} - w_{aa}) = 0$, from which a little algebra gives

$$\hat{p} = \frac{w_{Aa} - w_{aa}}{2w_{Aa} - w_{AA} - w_{aa}} = \frac{s_{aa}}{s_{AA} + s_{aa}} \tag{2.36}$$

where s_{AA} and s_{aa} are the selection coefficients against AA and aa homozygous genotypes, relative to a value of $w_{Aa} = 1$ for heterozygous genotypes.

The equilibrium in Equation 2.36 is globally stable, whereas those at $p = 0$ and $p = 1$ are now unstable, as indicated in Figure 2.12A, where the arrowheads show the direction of change in allele frequency. Figure 2.12B shows the course of change in $\bar{w}$ with overdominance, and it is of interest that the average fitness in the population is maximal when $p = \hat{p}$. Maximization of average fitness is a frequent outcome of selection in random-mating populations with constant fitnesses, but there are many exceptions to the rule when random mating does not occur, when fitnesses are not constant, or when more than one gene is involved (Ewens 1979; Curtsinger 1984). Note particularly that $\bar{w}$ is the average fitness *in* the population, not the average fitness *of* the population. The relative survivorships are relevant only to the differential mortality of the genotypes within a population at any given time. The average of the relative survivorships $\bar{w}$ has no necessary

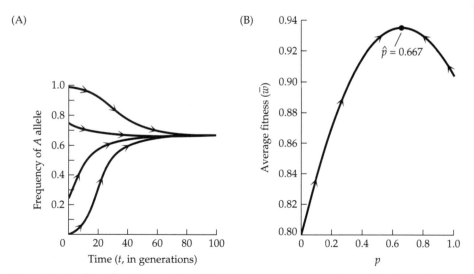

Figure 2.12 Overdominance. (A) Allele frequency converges to a globally stable equilibrium. (B) The average fitness in the population is maximized at equilibrium. The relative fitnesses are $w_{AA} = 0.9$, $w_{Aa} = 1$, and $w_{aa} = 0.8$.

relation to vernacular meanings of "fitness," such as competitive ability, population size, production of biomass, or evolutionary persistence (Haymer and Hartl 1983). The distinction is critical because, if the absolute mortality is high enough, a population can be going extinct even as selection is maximizing the average fitness in the population.

Although overdominance might seem to be a potent force for maintaining polymorphisms in natural populations, overdominance has been documented in only a few cases. A classic example is the sickle-cell anemia mutation which, when heterozygous, increases resistance to falciparum malaria but which, when homozygous, causes severe anemia. The relative viabilities in high-malaria regions in Africa have been estimated as $w_{AA} = 0.85$, $w_{Aa} = 1$, and $w_{aa} = 0$ (Allison 1964), where A represents the nonmutant allele and a the sickle-cell allele. Substitution into Equation 2.36 leads to the predicted equilibrium allele frequencies $\hat{p} = 0.87$ and $\hat{q} = 0.13$. These are reasonably close to the average allele frequencies observed in West Africa, but there is considerable variation among local populations.

The opposite of overdominance is **heterozygote inferiority,** which occurs when the heterozygote is the least fit ($w_{Aa} < w_{AA}$ and $w_{Aa} < w_{aa}$). A third equilibrium exists in this case also, and its value is given by Equation 2.36. However, in the case of heterozygote inferiority, the equilibrium is unstable, whereas the equilibria at $p = 0$ and $p = 1$ are both locally (but not globally) stable. An example of heterozygote inferiority is depicted in Figure 2.13A, where the arrows again denote the direction of change in allele frequency. If

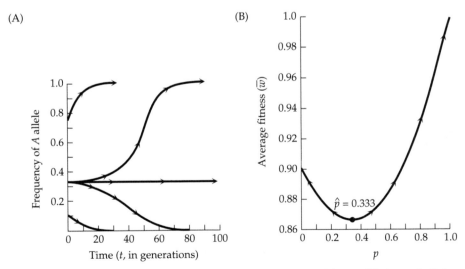

Figure 2.13 Heterozygote inferiority. (A) There is an unstable equilibrium of allele frequency. (B) The average fitness in the population is minimized at the unstable equilibrium. The relative fitnesses are $w_{AA} = 1$, $w_{Aa} = 0.8$, and $w_{aa} = 0.9$.

the initial allele frequency is exactly equal to the equilibrium value (in this example, $\hat{p} = 1/3$), then the allele frequency remains at that value. Otherwise, p goes to 0 or 1, depending on whether the initial allele frequency was less than or greater than the equilibrium value. Figure 2.13B plots the average fitness in the population. In this case, the unstable equilibrium at $\hat{p} = 1/3$ represents a *minimum* of average fitness.

Mutation–Selection Balance

Recall from Chapter 1 that outcrossing species typically contain a large amount of hidden genetic variability in the form of recessive or nearly recessive harmful alleles at low frequencies. This situation is to be expected. Selection cannot completely eliminate harmful alleles because of their continual creation through recurrent mutation. To be specific, suppose a is a harmful allele and that mutation of A to a occurs at the rate μ per generation. Because q, the frequency of a, remains small, reverse mutation of a to A can safely be ignored. The approximation for Δp in Equation 2.32 is still valid, except that a proportion $p\mu$ of A alleles mutate to a in the course of each generation. Therefore,

$$\Delta p = pqs\left[ph + q(1-h)\right] - p\mu \tag{2.37}$$

When selection is balanced by recurrent mutation, there is a globally stable equilibrium at the allele frequencies $\hat{p}$ and $\hat{q}$ that solve Equation 2.37 for $\Delta p = 0$. The two cases of greatest interest are:

1. **When the harmful allele is completely recessive ($h = 0$).** In this case the equilibrium equation becomes $pq^2s = \mu$, and since $\hat{p} \approx 1$, then

$$\hat{q} = \sqrt{\frac{\mu}{s}} \tag{2.38}$$

2. **When the harmful allele is partially dominant ($h > 0$).** Here the equilibrium equation is $pqhs + q^2s(1 - h) = \mu$, and since $\hat{p} \approx 1$, then $\hat{q}^2 \approx 0$, and so

$$\hat{q} = \frac{\mu}{hs} \tag{2.39}$$

Since Equation 2.38 involves the square root of the mutation rate, whereas Equation 2.39 involves the mutation rate itself, the equilibrium frequency $\hat{q}$ in Equation 2.38 is generally considerably larger than that in Equation 2.39. In other words, partial dominance even as small as 1% ($h = 0.01$) or 2%

($h = 0.02$) can have a significant effect on decreasing the equilibrium frequency of a harmful recessive allele maintained by recurrent mutation.

Huntington disease serves as an example of the use of Equation 2.39. The disease is a severe degenerative disorder of the neuromuscular system that typically appears after age 35. Although the disease itself results from a "dominant" gene, the fitnesses have been estimated as $w_{AA} = 1$, $w_{Aa} = 0.81$, and $w_{aa} = 0$, where A and a refer, respectively, to the nonmutant and mutant alleles (Reed and Neel 1959). Hence $s = 1$ and $h = 0.19$. In terms of neuromuscular degeneration, the Huntington allele is dominant. In terms of fitness, the allele is only partially dominant because of the late age of onset of the disease. Since we are dealing here with partial dominance, we could use Equation 2.39 to estimate $\hat{q}$ if μ were known, or alternatively to estimate μ if $\hat{q}$ were known. In the Michigan population in which the fitnesses were estimated, $\hat{q} = 5 \times 10^{-5}$. Assuming that the population is in equilibrium, we therefore have $\mu = \hat{q}hs = (5 \times 10^{-5})(0.19) = 9.5 \times 10^{-6}$. This example illustrates one of the common indirect methods for the estimation of mutation rates in humans.

Harmful recessive alleles that are too frequent to be explained by recurrent mutation are sometimes suspected of conferring a slight heterozygous advantage, but it is usually quite difficult to prove the hypothesis or to identify the source of selection. The case of cystic fibrosis is an example in which the frequency of the harmful allele is $q = 0.02$. In the absence of intensive medical care, the disease is lethal, so effectively $s = 1$. Assuming equilibrium according to Equation 2.38, we would have to postulate that $\mu = q^2 = 4 \times 10^{-4}$ to explain the allele frequency. This rate of mutation is far larger than can easily be explained. The alternative of overdominance is supported by the finding that the CFTR (cystic fibrosis transmembrane conductance regulator) is used by the agent of typhoid fever, *Salmonella typhi*, for passage through the intestinal mucosa to initiate infection. Heterozygous mice carrying the most common mutation in the human population (the *ΔF508* deletion allele) exhibited an 86% reduction in bacterial transfer through the mucosa, and homozygous mice blocked transfer altogether (Pier et al. 1998). The degree of heterozygous advantage needed to explain the allele frequency can be deduced from Equation 2.36, which implies that $\hat{q} = s_{AA}/(s_{AA} + s_{aa})$, where $s_{aa} = 1$ and s_{AA} is the selection coefficient against the nonmutant homozygous genotype, relative to a value of $w_{Aa} = 1$ for the heterozygous genotype. Substituting $\hat{q} = 0.02$ and $s_{aa} = 1$ yields $s_{AA} = 0.02$, which means that the heterozygous genotype may have about a 2% advantage in fitness due to increased typhoid fever resistance.

More Complex Modes of Selection

We have considered only the simplest type of selection involving two alleles of a single gene with constant fitnesses determined by differences in viability. In this section we briefly define some of the more complex models to illustrate some of the issues. Detailed discussion of the models is beyond the scope of this book, but more information and references can be found in Hartl and Clark (1997). Complexities arise when the fitnesses of the genotypes or changes in allele frequency depend on:

- Multiple alleles, because there may be multiple equilibria
- Multiple loci and gene interaction (epistasis or epistatic selection)
- Allele frequencies or genotype frequencies (frequency-dependent selection)
- Population size (density-dependent selection)
- Both genotypes present in each mating pair (fecundity selection)
- Differential survival or function of gametes (gametic selection)
- Non-Mendelian segregation in heterozygous genotypes (meiotic drive)
- Age or developmental stage of the organism (overlapping generations)
- Environmental fluctuations (heterogeneous environments)
- Rarity of a given allele or genotype (rare-allele advantage, diversifying selection)
- Sex of the organism (sex-limited selection)
- Chromosomal location of the gene (X-linked or Y-linked selection)
- Cellular location of the gene (mitochondrial or chloroplast selection)
- Success in attracting mates (sexual selection)
- Behaviors enhancing the fitness of genetically related individuals (kin selection)
- Differential extinction and recolonization of subpopulations (interdeme selection)

RANDOM GENETIC DRIFT

The term **random genetic drift (RGD)** refers to fluctuations in allele frequency that occur by chance, particularly in small populations, as a result of random sampling among gametes. The fundamental reason for RGD is illustrated in Figure 2.3. A population of N diploid zygotes arises from a sample of $2N$ gametes chosen from an essentially infinite pool of gametes. Because the population size is finite, two alleles present in the zygotes of any generation may be exact replicas of a single allele present in breeding adults in the previous generation. These alleles are identical by descent (IBD). As time goes on the probability of IBD increases as specified in Equation 2.9, where F_t is the fixation index F_{ST} in generation t.

Diffusion Approximations

Suppose now that the population in Figure 2.3 is actually one of an infinite number of replicate subpopulations initiated with the same allele frequencies $\bar{p}$ and $\bar{q}$ and allowed to evolve independently with no migration. What happens in the metapopulation as a whole? As we have seen, the genotype frequencies in any subpopulation are in HWE because of random mating within each subpopulation. But the genotype frequencies in the metapopulation are given by Equation 2.24 and are not in HWE. These equations do, however, indicate that the allele frequencies in the metapopulation remain constant at $\bar{p}$ and $\bar{q}$. In other words, even as the allele frequencies diverge among the subpopulations due to RGD, the average allele frequencies among the subpopulations remain constant at $\bar{p}$ and $\bar{q}$. At any time t the variance in allele frequency among the subpopulations is given by Equation 2.23 as $\sigma^2 = \bar{p}\,\bar{q}\,F_{ST}$.

For some purposes it is not enough to know the mean and variance of allele frequencies among the subpopulations. We wish to know the statistical distribution of allele frequencies among the subpopulations, which changes through time. An experimental example is shown in Figure 2.14,

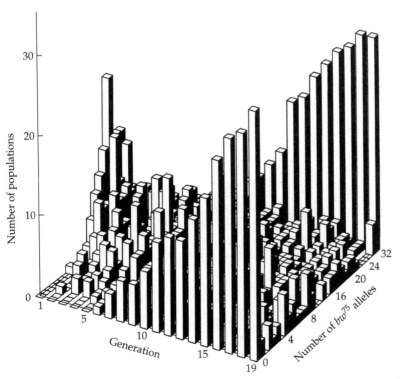

Figure 2.14 Random genetic drift in 107 experimental subpopulations of *Drosophila melanogaster*. (Data from Buri 1956.)

which records the history of 19 generations of random genetic drift in 107 subpopulations of *Drosophila melanogaster* (Buri 1956). Each population was initiated with $N = 16$ bw^{75}/bw heterozygous flies ($bw = brown\ eyes$) and maintained at a constant size of 16 by randomly choosing 8 males and 8 females as the breeding population in each generation. The plot in Figure 2.14 gives the number of populations having $0, 1, 2, \ldots, 32$ bw^{75} alleles in each generation. As the experiment progresses, the initially humped distribution of allele frequency gradually becomes flat as populations fixed for bw^{75} or bw begin to pile up. By about 18 generations, half the populations are fixed for one allele or the other, and among the unfixed populations, the distribution of allele frequencies is essentially flat.

The changing distribution of allele frequencies shown in Figure 2.14 is a good match to the theoretical expectation based on repeated application of the binomial distribution (Figure 2.15). Consider a particular subpopulation with allele frequency p of A. In the next generation, the number i of copies of the A allele in the subpopulation may be $i = 0, 1, 2, \ldots, 2N$, each possible

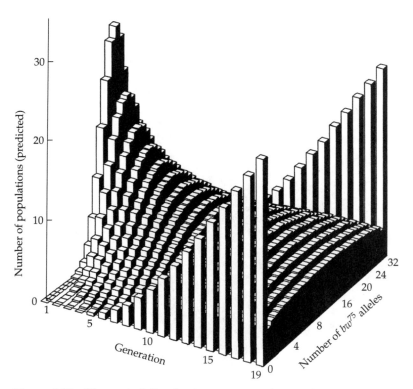

Figure 2.15 Theoretical distribution corresponding to the experiment in Figure 2.14. (Data from Buri 1956.)

outcome having a different probability given by successive terms in the binomial expansion

$$\left(p\,A+q\,a\right)^{2N} = \sum_{i=0}^{2N} \frac{(2N)!}{i!(2N-i)!}\,p^i q^{2N-i} \tag{2.40}$$

where $i!$ is the product of all integers up to and including i, but by definition $0! = 1$. If the expansion in Equation 2.40 is carried out for each individual subpopulation in any generation, and then the terms corresponding to $i = 0$, $1, 2, \ldots, 2N$ are summed over all subpopulations, the result is the distribution of allele frequencies in the next generation. Repeating this process over and over yields the theoretical distribution in Figure 2.15.

Although Equation 2.40 is exact, there is a diffusion approximation to RGD that is somewhat more tractable (Wright 1945; Kimura 1957, 1964). The diffusion approximation assumes that RGD disperses allele frequencies among subpopulations in a manner analogous to heat diffusing through a metal rod or tiny particles diffusing under Brownian motion (Kolmogorov 1931). The idea is to assume that the subpopulations are large enough that the allele frequencies change smoothly through time, not in large jumps. Then the statistical distribution of allele frequencies at any time is a continuous function that we may denote $\phi(p, x; t)$, where p is the initial allele frequency among all the subpopulations, x is the current frequency ($0 < x < 1$), and t is the generation number. For any time t, the function ϕ is a smooth, continuous function approximating the histogram of allele frequencies in Figure 2.15 in generation t, except that ϕ pertains only to the distribution of allele frequencies among subpopulations still segregating for A and a. Figure 2.16A shows $\phi(0.5, x; t)$, which corresponds to various time slices through Figure 2.15, and Figure 2.16B shows $\phi(0.1, x; t)$.

To explain the meaning of the diffusion equations, let us for the moment interpret p as the number of A alleles present in a subpopulation initially and x as the number present at time t, with $1 < p, x < 2N$. We will also interpret $\phi(p, x; t)$ as the proportion of subpopulations that have changed from having p copies at time 0 to having x copies at time t. These interpretations are not rigorous from a mathematical standpoint, but we use them merely to explain the diffusion equations not derive them.

To find an equation for $\phi(p, x; t)$ we may first look backward in time to the beginning of the process and consider what may have happened in the very first generation. The bookkeeping is shown in Table 2.2. We assume that the state of each subpopulation changes slowly enough that in any one generation the number of copies of an allele (the "state" of the subpopulation) either stays the same, increases by 1, or decreases by 1. There are two reasons why the state could change. One is RGD, the other is a systematic force, denoted

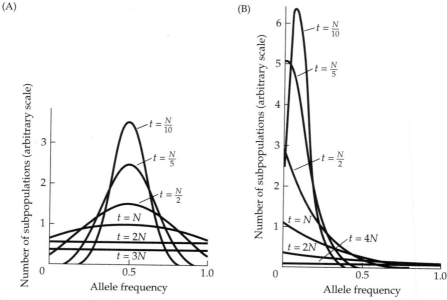

Figure 2.16 Continuous approximations to the distribution of allele frequency among subpopulations, excluding those fixed for A or for a. (A) When initially $p = 0.5$, corresponding to time-slices through Figure 2.15. (B) When initially $p = 0.1$. (From Kimura 1955.)

SF, which might include mutation or selection. We will assume that A is the favored allele and define $M(p)$ as the probability that p increases from p to $p + 1$ because of the SF. The force of RGD is measured by the probability $V(p)$ that p changes, because of RGD, either to $p - 1$ (probability $V(p)/2$) or to

Table 2.2 Routes from p to x in t generations, looking backward to the first generation

Possibilities for change in first generation	Probability of first-generation change	Probability of changing to x in remaining $t - 1$ generations
$p \to p + 1$ because of SF	$M(p)$	$\phi(p + 1, x; t - 1)$
$p \to p + 1$ because of RGD	$V(p)/2$	$\phi(p + 1, x; t - 1)$
$p \to p - 1$ because of RGD	$V(p)/2$	$\phi(p - 1, x; t - 1)$
p remains at p	$1 - M(p) - V(p)$	$\phi(p, x; t - 1)$

$$\phi(p, x; t) - \phi(p, x; t - 1) = M(p)[\phi(p + 1, x; t - 1) - \phi(p, x; t - 1)]$$

$$+ \frac{V(p)}{2}\{[\phi(p + 1, x; t - 1) - \phi(p, x; t - 1)] - [\phi(p, x; t - 1) - \phi(p - 1, x; t - 1)]\}$$

$p + 1$ (probability $V(p)/2$). Therefore, in the first generation, p can change to $p + 1$ with probability $M(p) + V(p)/2$, to $p - 1$ with probability $V(p)/2$, or it can remain at p with probability $1 - M(p) - V(p)$. For a subpopulation to be in state x in generation t, the state must change from whatever it was after the first generation to the state x in the last $t - 1$ generations. The probabilities of changing from $p + 1$ to x, or from p to x, or from $p - 1$ to x in $t - 1$ generations are, respectively, $\phi(p + 1, x; t - 1)$, $\phi(p, x; t - 1)$, and $\phi(p - 1, x; t - 1)$. The overall proportion of subpopulations in state x at time t is $\phi(p, x; t)$, which is obtained by summing the products of columns 2 and 3 in Table 2.2. After some rearrangement the difference equation at the bottom results, which gives the change in ϕ with respect to time expressed in terms of the change in ϕ with respect to the initial state. When p and x are reconsidered as allele frequencies (not whole numbers) and ϕ reconsidered as a statistical distribution (not a probability), then the terms in the difference equation at the bottom of Table 2.2 converge in the limit to the terms of the **Kolmogorov backward equation**

$$\frac{\partial \phi(p,x;t)}{\partial t} = M(p)\frac{\partial \phi(p,x;t)}{\partial p} + \frac{V(p)}{2}\frac{\partial^2 \phi(p,x;t)}{\partial p^2} \qquad (2.41)$$

Some of the implications of this equation for population genetics are summarized in the next section.

A second type of diffusion equation can be obtained by considering $\phi(p, x; t)$ and asking how the distribution changes as we go forward in time for one generation. The bookkeeping in this case is shown in Table 2.3. Because changes in state are limited to $+1$ or -1, a subpopulation can be in state x at time $t + 1$ only if it was in state $x + 1$, x, or $x - 1$ at time t, and these have probabilities $\phi(p, x + 1; t)$, $\phi(p, x; t)$, and $\phi(p, x - 1; t)$, respectively. A sub-

Table 2.3 Routes from p to x in $t + 1$ generations, looking forward from time t

Possibilities for frequency after t generations	Probability	Possibilities to go to x in next generation	Probability
$x - 1$	$\phi(p, x - 1; t)$	$x - 1 \rightarrow x$ because of SF	$M(x - 1)$
	$\phi(p, x - 1; t)$	$x - 1 \rightarrow x$ because of RGD	$V(x - 1)/2$
$x + 1$	$\phi(p, x + 1; t)$	$x + 1 \rightarrow x$ because of RGD	$V(x + 1)/2$
x	$\phi(p, x; t)$	x remains at x	$1 - M(x) - V(x)$

$$\phi(p, x; t + 1) - \phi(p, x; t) = -[M(x)\phi(p, x; t) - M(x - 1)\phi(p, x - 1; t)]$$

$$+ \frac{1}{2}\{[V(x + 1)\phi(p, x + 1; t) - V(x)\phi(p, x; t)] - [V(x)\phi(p, x; t) - V(x - 1)\phi(p, x - 1; t)]\}$$

population in state $x - 1$ can change to state x with probability $M(x - 1) + V(x - 1)/2$ according to whether it was pushed by the SF or changed randomly by RGD. A subpopulation in state $x + 1$ can change to state x with probability $V(x + 1)/2$ due to RGD. And a subpopulation in state x can remain in state x with probability $1 - M(x) - V(x)$. The required function $\phi(p, x; t + 1)$ is obtained by summing the products of rows 2 and 4 in Table 2.3, yielding the difference equation at the bottom. In this case, the expression relates changes in ϕ with respect to time to changes in ϕ with respect to its current state. When p and x are reconsidered as allele frequencies and ϕ as a statistical distribution, then the terms in the difference equation at the bottom of Table 2.3 converge in the limit to the terms of the **Kolmogorov forward equation**

$$\frac{\partial \phi(p,x;t)}{\partial t} = -\frac{\partial \left[M(x)\phi(p,x;t)\right]}{\partial x} + \frac{1}{2}\frac{\partial^2 \left[V(x)\phi(p,x;t)\right]}{\partial x^2} \tag{2.42}$$

The solutions to this equation are the curves plotted in Figure 2.16.

Although we have explained the underlying logic of the diffusion equations, we have not yet specified $M(p)$, $V(p)$, $M(x)$, or $V(x)$ in terms that have any relation to population genetics. $M(p)$ is another symbol for the change in allele frequency that occurs in one generation due to any systematic force such as mutation, migration, or selection. We have variously symbolized $M(p)$ as $p_{t+1} - p_t$ (Equations 2.4 for mutation and 2.13 for migration), $p' - p$ (Equations 2.25 and 2.29 for selection), and Δp (Equations 2.31 and 2.32 for selection). $M(x)$ is the same as $M(p)$ but the allele frequency is symbolized as x instead of p. The term for RGD also has a straightforward biological interpretation. $V(p)$ is the variance in allele frequency after one generation of binomial sampling of $2N$ alleles according to the expansion in Equation in 2.40; hence, $V(p) = p(1 - p)/(2N)$. Similarly, $V(x) = x(1 - x)/(2N)$.

Probability of Fixation and Time to Fixation

Analysis of the backward equation (Equation 2.41) has led to many important insights in population genetics. Mathematical details can be found in Crow and Kimura (1970), Kimura and Ohta (1971), Kimura (1983), and Ewens (1979). The approach is to assume that time is so advanced that ϕ is no longer changing, so that the left hand side can be set to 0. Then one can set $x = 1$ to obtain the probability that an allele with initial frequency p eventually becomes fixed in a particular subpopulation. The **fixation** of A means that the allele frequency of A goes from an initial frequency of p to final frequency of 1. The most important special case is that of a new mutant allele present initially in a single copy so that $p = 1/(2N)$. In the formulas that follow, we give the results for any initial frequency p but also, after the right arrow, for the case $p = 1/(2N)$.

In the case of an unselected or **neutral allele** in a finite population,

$$\text{Pr}\{\text{fixation}\} = p \to \frac{1}{2N} \text{ when } p = \frac{1}{2N} \qquad (2.43)$$

This means that the probability of ultimate fixation of a neutral allele is equal to its initial frequency in the subpopulation. This result also follows intuitively from Equation 2.24, where we use $\bar{p}$ for the initial frequency instead of p. Since the average allele frequency among all subpopulations remains $\bar{p}$, then at a time when all subpopulations have become fixed, the proportion that are fixed for A must equal $\bar{p}$.

In the case of selection, let s be the selective advantage of the heterozygous genotype and assume additivity, so the relative fitnesses are $1 + 2s$ for AA and $1 + s$ for Aa, relative to a value of 1 for aa. Then

$$\text{Pr}\{\text{fixation}\} = \frac{1 - e^{-4Nsp}}{1 - e^{-4Ns}} \to \frac{1 - e^{-2s}}{1 - e^{-4Ns}} \text{ when } p = \frac{1}{2N} \qquad (2.44)$$

If $Ns \ll -1$ (the double inequality means "much less than"), or $Ns \gg +1$, then the fate of a new mutation is determined almost exclusively by selection. On the other hand, for $-1 < Ns < 1$, both selection and RGD play a role. The case $p = 1/(2N)$ can be simplified further when s is small and N is large, in which case the probability of fixation $\approx 2s$. This means that the fixation of a favorable allele is by no means certain, even in a large population. For example, in a population of size 10^6, a new favorable mutation that confers a 1% fitness advantage in heterozygotes has only a 2% chance of eventually becoming fixed. On the other hand, Equation 2.43 implies that a neutral allele in the same population has only a 0.00005% chance of becoming fixed. It is worth noting that Equation 2.44 also applies to a detrimental mutation, which means $s < 0$ (in which case A is the disfavored allele). A detrimental mutation has little chance of becoming fixed unless $|Ns|$ is quite close to 0 (Ohta 1973, 1992).

Other important results that can be derived from the Kolmogorov backward equation are the average times to fixation of alleles that eventually are fixed and the average time to loss of alleles that eventually are lost. Analytical results are not possible except in the neutral case (Ewens 1979), but in the neutral case,

$$\text{Avg time to fixation} = -\frac{1}{p}\left[4N(1-p)\ln(1-p)\right] \rightarrow \approx 4N \text{ when } p = \frac{1}{2N} \quad (2.45)$$

$$\text{Avg time to loss} = -4N\frac{p}{1-p}\ln p \rightarrow \approx 2\ln 2N \text{ when } p = \frac{1}{2N} \quad (2.46)$$

The main implication is that, for new mutant alleles that are selectively neutral, for which $p = 1/(2N)$, those that are destined to become fixed take a long time to be fixed, but those that are destined to be lost are lost quickly. In a population of size 10^6, for example, the average time to fixation is 4 million generations, but the average time to loss is 29 generations

Effective Population *Number*

The theoretical model of RGD fits the experimental data in Figure 2.14 reasonably well, except in one respect. The average heterozygosity in the subpopulations is supposed to decrease according to

$$H_t = 2\bar{p}\bar{q}(1-F_{ST}) = 2\bar{p}\bar{q}\left(1-\frac{1}{2N}\right)^t \quad (2.47)$$

which is obtained by combining Equation 2.24 with Equation 2.9 and setting $F_0 = 0$. As noted, $N = 16$ in the experimental subpopulations, and the expected decrease in H_t is shown in Figure 2.17. The fit of the actual points is not very impressive, especially for the later generations. However, the pattern of decrease in H_t does follow the theoretical prediction for the value $N = 9$. In these subpopulations, the main reason for the discrepancy is that the distribution of offspring number has a variance greater than the mean offspring number. This violates one of the underlying assumptions of the theory, which is that every individual has an equal likelihood of leaving offspring. (Technically, the theory assumes a Poisson distribution of offspring number.) Nevertheless, the theory fits with $N = 9$, and so we call $N = 9$ the **effective number** in the experimental subpopulations, usually symbolized N_e. More precisely, we should call $N_e = 9$ the **inbreeding effective number** because $N_e = 9$ is the size of theoretically ideal subpopulations that undergo the same rate of increase in F (decrease in H_t) as the actual subpopulations. The effective size of a population is almost always smaller than the actual size.

In considerations of RGD, the effective population number is usually of greater interest than the actual population number. There are three important instances in which the two quantities are related in a reasonably simple manner.

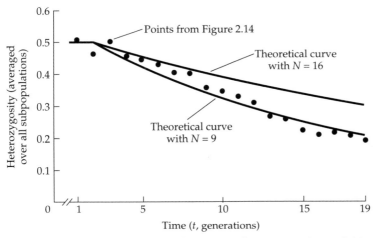

Figure 2.17 Theoretical curves for average heterozygosity with $N = 16$ or $N = 9$, along with actual values from the experiment in Figure 2.14.

1. **Variable population size.** Imagine a subpopulation that is ideal in all respects except that its number changes from generation to generation. Suppose, to be precise, that its effective number is N_1 in generation 1, N_2 in generation 2, and so forth. In this situation, the overall decrease in H_t is given by Equation 2.47 with N replaced by N_e, where

$$\frac{1}{N_e} = \frac{1}{t}\left(\frac{1}{N_1} + \frac{1}{N_2} + \cdots + \frac{1}{N_t}\right) \tag{2.48}$$

Equation 2.48 says that the average effective population number is calculated as the reciprocal of the average of the reciprocals. This is a special sort of average called the harmonic mean, which tends to be dominated by the smallest terms. Suppose, for example, that $N_1 = 1000$, $N_2 = 10$, and $N_3 = 1000$ in a population that underwent a severe temporary reduction in population size (a **bottleneck**) in generation 2. Then $1/N_e = (1/3)$ $(1/1000 + 1/10 + 1/1000) = 0.034$. The average effective number over the three-generation period is only $N_e = 29.4$, whereas the average actual number is $(1/3)(1000 + 10 + 1000) = 670$. A severe population bottleneck often occurs in nature when a small group of emigrants from an established subpopulation founds a new subpopulation. The random genetic drift accompanying such a founder event is known as a **founder effect**.

2. **Unequal sex ratio.** An inequality in the sex ratio creates a peculiar sort of "bottleneck" because half of the alleles in any generation must come from

each sex no matter how few individuals of the minority sex there are. If there are N_m males and N_f females, then

$$N_e = \frac{4N_mN_f}{N_m + N_f} \tag{2.49}$$

Not surprisingly, N_e equals two times the harmonic mean of N_m and N_f. To take an example from wildlife management, suppose that deer hunting is permitted to a level at which the number of surviving males is one-tenth the number of surviving females. Then $N_m = 0.1 \times N_f$ and Equation 2.49 implies that N_e is reduced to about one-third of the actual number (N_e is $0.36 \times N_m$, whereas the actual size is $11 \times N_m$).

3. **Uniform population dispersion.** For a population spread out uniformly in two dimensions the effective size depends on (1) the number of breeding individuals per unit area, usually denoted by the symbol δ, and (2) the amount of dispersion between an individual's own birthplace and that of its offspring. If dispersion follows a normal, bell-shaped curve in both dimensions with standard deviation σ, then 39% of all individuals have their offspring within a circle of radius σ centered at their own birthplace, 87% have their offspring within a circle of radius 2σ, and 99% have their offspring within a circle of radius 3σ. In terms of δ and σ, the effective size of the population (often called **neighborhood size** in this context) is given by

$$N_e = 4\pi\delta\sigma^2 \tag{2.50}$$

where $\pi = 3.14159$.

Equation 2.50 can be applied to data on the abundant prairie deer mouse *Peromyscus maniculatus*. In a large area in Southern Michigan, Dice and Howard (1951) estimated the density of breeding individuals δ to be in the range of 5–7 per hectare (one hectare equals 10,000 square meters or about 2.5 acres). By following the movement of marked animals from birth to breeding sites, they estimated σ as 114 meters, yielding $\sigma^2 = 1.3$ hectares. With these parameters, Equation 2.50 yields an the effective size in the range 81.7–114.4, which is perhaps surprisingly small for such an abundant animal.

FURTHER READINGS

Cavalli-Sforza, L. L. 1996. *History and Geography of Human Genes.* Princeton University Press, Princeton, NJ.

Cavalli-Sforza, L. L. and W. F. Bodmer. 1971. *The Genetics of Human Populations.* W. H. Freeman and Co., San Francisco.

Charlesworth, B. 1994. *Evolution in Age-Structured Populations,* 2nd Ed. Cambridge University Press, Cambridge.

Crow, J. F. and M. Kimura. 1970. *An Introduction to Population Genetics Theory.* Harper & Row, New York.

Ewens, W. J. 1979. *Mathematical Population Genetics.* Springer-Verlag, Berlin.

Futuyma, D. J. and M. Slatkin. 1983. *Coevolution.* Sinauer Associates, Sunderland, MA.

Gillespie, J. H. 1991. *The Causes of Molecular Evolution.* Oxford University Press, Oxford.

Golding, B. (ed.). 1994. *Non-Neutral Evolution: Theories and Molecular Data.* Chapman and Hall, New York.

Graur, D. and W.-H. Li. 2000. *Fundamentals of Molecular Evolution,* 2nd Ed. Sinauer Associates, Sunderland, MA.

Hartl, D. L. and A. C. Clark. 1997. *Principles of Population Genetics,* 3rd Ed. Sinauer Associates, Sunderland, MA.

Hamrick, J. L. (ed.). 1995. *Conservation Genetics: Case Histories from Nature.* Chapman and Hall, New York.

Jacquard, A. 1978. *Genetics of Human Populations.* Trans. by D. M. Yermanos. Jones & Bartlett, Boston.

Kimura, M. 1983. *The Neutral Theory of Molecular Evolution.* Cambridge University Press, Cambridge.

Kimura, M. and T. Ohta. 1971. *Theoretical Aspects of Population Genetics.* Princeton University Press, Princeton, NJ.

Lewontin, R. C. 1974. *The Genetic Basis of Evolutionary Change.* Columbia University Press, New York.

Manly, B. F. J. 1985. *The Statistics of Natural Selection on Animal Populations.* Chapman and Hall, London.

Mitton, J. F. 1997. *Selection in Natural Populations.* Oxford University Press, New York.

Nagylaki, T. 1977. *Selection in One- and Two-Locus Systems,* Lecture Notes in Biomathematics, Vol. 15. Springer-Verlag, Berlin.

Nei, M. 1987. *Molecular Evolutionary Genetics.* Columbia University Press, New York.

Nei, M. and S. Kumar. 2000. *Molecular Evolution and Phylogenetics.* Oxford University Press, New York.

Ohta, T. and K. Aoki. 1985. *Population Genetics and Molecular Evolution.* Springer-Verlag, New York.

Roughgarden, J. 1995. *Theory of Population Genetics and Evolutionary Ecology: An Introduction.* Prentice Hall, Paramus, NJ.

Selander, R. K., A. G. Clark and T. S. Whittam (eds.). 1991. *Evolution at the Molecular Level.* Sinauer Associates, Sunderland, MA.

Weir, B. S. 1996. *Genetic Data Analysis II.* Sinauer Associates, Sunderland, MA.

Weir, B. S., E. J. Eisen, M. M. Goodman and G. Namkoong (eds.). 1988. *Proceedings of the Second International Conference on Quantitative Genetics.* Sinauer Associates, Sunderland, MA.

PROBLEMS

2.1 The *mariner* transposable element in *Drosophila mauritiana* results in the spontaneous deletion of a target sequence at a frequency of approximately 1% per generation (Hartl et al. 1997). In a population in which the target sequence is fixed (homozygous), how many generations would be required for the expected frequency of flies with a homozygous deletion of the target element to exceed 5%? Assume that the population is large, that mating is random, and that deletion of the target sequence does not affect survival or reproduction.

2.2 Stocker (1949) studied a case in the bacterium *Salmonella typhimurium* in which the mutation rates were sufficiently large that the equations for mutational equilibria could be tested. The gene in question controls a protein component of the cellular flagella. There are two alleles, which we can call A and a. The mutation rate from A to a was estimated as $\mu = 8.6 \times 10^{-4}$ per generation, and that of a to A as $v = 4.7 \times 10^{-3}$ per generation. (These mutation rates are orders of magnitude larger than typically observed for other genes. The reason is that the change from A to a and back again does not involve mutation in the conventional sense, but results from intrachromosomal recombination.) In cultures initially established with a frequency of A at $p_0 = 0$, Stocker found that the frequency increased to $p = 0.16$ after 30 generations and to $p = 0.85$ after 700 generations. In cultures initiated with $p_0 = 1$, the frequency decreased to 0.88 after 388 generations and to 0.86 after 700 generations. How do these values agree with those calculated from Equation 2.5 using the estimated mutation rates? What is the predicted equilibrium frequency of A?

2.3 The *gnd* gene in *E. coli* encodes the enzyme 6-phosphogluconate dehydrogenase (6PGD), which is used in the metabolism of gluconate but not ribose. When otherwise genetically identical strains containing the naturally occurring alleles *gnd(RM77C)* or *gnd(RM43A)* were placed in competition on gluconate or ribose, the data below were obtained, in which p denotes the frequency of the strain containing *gnd(RM43A)* (Dykhuizen and Hartl 1980; Hartl and Dykhuizen 1981). Estimate the fitness of the strain with *gnd(RM77C)* relative to that with *gnd(RM43A)* under the two growth conditions.

Genotype of competing strains	Growth medium	p_0	p_{35}
gnd(RM43A) vs gnd(RM77C)	Gluconate	0.455	0.898
gnd(RM43A) vs gnd(RM77C)	Ribose	0.594	0.587

2.4 In *D. melanogaster, curly wings* is due to a dominant allele *Cy* that is lethal when homozygous. A population is established with an initial frequency of *Cy* equal to 0.168. Calculate the expected frequency in the next generation, assuming:

a. that the relative fitness of +/+ : *Cy*/+ is 1 : 1.
b. that the relative fitness of +/+ : *Cy*/+ is 1 : 0.5.

2.5 In the evolution of industrial melanism in *Biston betularia* (see Problem 1.5), the allele resulting in black body coloration may be considered a favored dominant. In this species, the frequency of melanic moths increased from a value of 1% in 1848 to a value of 95% in 1898. The species has one generation per year.

    a. Estimate the approximate value of the selection coefficient s against non-melanics that would be necessary to account for the change in frequency of the melanic forms.

    b. How many generations would be required for the same change in frequency of melanic forms in a hypothetical case in which the allele for melanism is recessive, assuming the same value of s against nonmelanics?

2.6   An extensively studied isolated colony of the moth *Panaxia dominula* contained a mutant allele affecting color pattern. The frequency of the mutant allele declined steadily over the period 1939–1968 in agreement with the theoretical prediction for a deleterious additive allele with $s = 0.20$. The species has one generation per year. In 1965 the estimated frequency of the mutant allele was 0.008. Estimate the frequency of the mutant allele in 1950 and in 1940.

2.7   Experimental populations of *Drosophila pseudoobscura* were established and periodically treated with weak doses of the insecticide DDT (Anderson et al. 1968). One population was initially polymorphic for five different inversions in the third chromosome, in approximately equal frequencies. After 13 generations, three of the inversions had disappeared from the population. The two that remained were *Standard* (*ST*) and *Arrowhead* (*AR*). Changes in frequency of each inversion were monitored, and from the values for the first nine generations the relative fitnesses of *ST/ST, ST/AR,* and *AR/AR* genotypes were estimated as 0.47, 1.0, and 0.62, respectively. Because the inversions yield almost no recombinant gametes, each type can be considered as an "allele." What equilibrium frequency of *ST* is predicted? What equilibrium value of $\bar{w}$ is predicted?

2.8   Warfarin is a blood anticoagulant used for rat control during and after World War II. Initially highly successful, the effectiveness of the rodenticide gradually diminished due to the evolution of resistance among some target populations. Among Norway rats in Great Britain, resistance results from an otherwise harmful mutation *WARF-R* in a gene in which the normal, sensitive allele may be denoted *WARF-S*. In the absence of warfarin, the relative fitnesses of *SS, SR,* and *RR* genotypes have been estimated as 1.00, 0.77, and 0.46 respectively. In the presence of warfarin, the relative fitnesses have been estimated as 0.68, 1.00, and 0.37, respectively (May 1985).

    a. Calculate the equilibrium frequency $\hat{q}$ of the resistance allele in the presence of warfarin.

    b. Noting that, in the absence of warfarin, *WARF-R* and *WARF-S* are very nearly additive in their effects on fitness, estimate the approximate number of generations required for the allele frequency of *WARF-R* to decrease from $\hat{q}$ to 0.01 in the absence of the anticoagulant.

2.9   A small amount of dominance can have a major effect in reducing the equilibrium frequency of a harmful allele. To confirm this for yourself, imagine an allele that is lethal when homozygous ($s = 1$) in a population of *Drosophila*.

Suppose that the allele is maintained by selection-mutation balance with $\mu = 5 \times 10^{-6}$. Calculate the equilibrium frequency of the allele in the cases:

a. Complete recessive.
b. Partial dominance with $h = 0.025$.

2.10   From four large chicken coops near Ramona, California, Selander and Yang (1969) trapped wild mice (*Mus musculus*) and carried out an electrophoretic study of a large number of proteins, including hexose-6-phosphate dehydrogenase, NADP–isocitrate dehydrogenase, and hemoglobin. Estimated values of F_{ST} for these genes were 0.10, 0.16, and 0.11, respectively; average $F_{ST} = 0.12$. Assuming that the mouse populations in each barn are about the same size (N), how long would it take for random genetic drift to result in a value of $F_{ST} = 0.12$ in the ideal (and undoubtedly oversimplified) case when migration between barns does not occur, assuming that $N = 20$? How long would it take when $N = 100$?

2.11   The neotropical giant toad *Bufo marinus* became established in Hawaii and Australia after the release of a large number of individuals in the 1930s. Allele frequencies of 10 polymorphic enzyme genes were studied in individuals from 10 local populations in Hawaii and 5 local populations in Australia (Easteal 1985). Estimated values of F_{ST} were 0.056 in Hawaiian populations, 0.063 in Australian populations, and 0.059 combined. The species has one generation per year, and there have been approximately 45 years since the introductions. Use these data to estimate the effective population number in the Hawaiian, Australian, and combined populations.

2.12   A highly isolated colony of the moth *Panaxia dominula* near Oxford, England, has been intensively studied by Ford and collaborators over the period 1928–1968 (Ford and Sheppard 1969). This species has one generation per year, and estimates of population size were carried out yearly beginning in 1941. For the years 1950 to 1961, inclusive, estimates of population size were as follows:

1950 4,100	1951 2,250	1952 6,000	1953 8,000
1954 11,000	1955 2,000	1956 11,000	1957 16,000
1958 15,000	1959 7,000	1960 2,500	1961 1,400

Assuming that the actual size of the population in any year equals the effective size in that year, estimate the average effective number over the entire 12-year period.

2.13   A dairy farmer has a herd consisting of 200 cows and 2 bulls. What is the effective size of the population?

2.14   In the Mohave desert, local populations of the diminutive annual plant *Linanthus parryae* ("desert snow") are polymorphic for white versus blue flowers. Blue flowers result from homozygosity for a recessive allele. The geographical distribution of the frequency q of the recessive allele across a region of the Mohave desert is shown in the accompanying illustration.

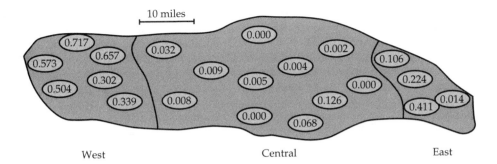

Each allele frequency is based on an examination of approximately 4000 plants over an area of about 30 square miles (Epling and Dobzhansky 1942). The highest frequencies of the blue-flower allele are largely concentrated at the west and east ends of the region in question. Treat each of the three regions as a single random-mating unit in HWE for the flower-color alleles. Estimate the average allele frequency in each region and in the population as a whole. From these data:

a. Estimate H_S and H_T for the flower-color gene.
b. Estimate F_{ST} for the flower-color gene.

2.15 Use Equations 2.38 and 2.39 to show that, at equilibrium for mutation–selection balance, the average relative fitness in the population is decreased from its maximum value by the amount μ (when a deleterious allele is completely recessive) or 2μ (when a deleterious allele shows partial dominance). Called the Haldane-Muller principle, this result shows that the total harm caused by a mutation (measured by the reduction in average fitness) depends on the rate of mutation only and (except for a factor of 2) is independent of the harmfulness or degree of dominance of any particular mutant allele.

2.16 Suppose that the selection intensity in a haploid population varies from generation to generation and that in generation i the relative fitnesses of $A : a$ are $1 : w_i$.

a. Show that $p_t/q_t = p_0/(q_0 \times w_0 w_1 \ldots w_{t-1})$.
b. If this is written as $p_t/q_t = p_0/(q_0 \times w^t)$, how can w be interpreted?

2.17 Use Equation 2.29 to show that the allele frequency of a recessive lethal in generation t, q_t, is given by $q_t = q_0/(1 + tq_0)$. (Hint: Find an expression for $1/q_t$ in terms of $1/q_{t-1}$ and solve by the method of successive substitutions.)

2.18 Show that a random-mating diploid population with relative fitnesses $1 : 1 - s : (1 - s)^2$ for AA, Aa, and aa has the same change in allele frequency as a haploid population with fitnesses $1 : 1 - s$ of A and a.

2.19 In an experimental population of *D. melanogaster* containing a segregation distorter chromosome known as *SD* (see Problem 1.19), the equilibrium frequency of the *SD* chromosome was approximately 0.125, and the segregation ratio in *SD* heterozygotes was about $k = 0.75$ (Hiraizumi et al. 1960). The *SD* chromosome is homozygous lethal in both sexes. The equilibrium between viability selection and segregation distortion in this case can be shown to be a wildtype allele frequency of $\hat{p} = 2(k-1)w_{12}/(1-2w_{12})$, where w_{12} is the fitness of the heterozygous SD genotype relative to homozygous wildtype. Use the equilibrium equation to estimate the approximate value of w_{12} consistent with the value of $\hat{p}$ in the experimental population.

2.20 For reversible mutation in a finite population, $M(x) = x' - x = (1-x)v - x\mu$ from Equation 2.3, and $V(x) = x(1-x)/2N$ from Equation 2.40. Show that $\phi(x) = Cx^{4Nv-1}(1-x)^{4N\mu-1}$ is an equilibrium solution to the Kolmogorov forward equation (Equation 2.42), where C is a constant chosen so that the integral of $\phi(x)$ from $x = 0$ to $x = 1$ equals 1.

Solutions to the problems, worked out in full, can be found at the website www.sinauer.com/hartl/html

CHAPTER 3

Molecular Population Genetics

Once upon a time population genetics was a largely theoretical subject. Its focus was on the relations between population structure, mating systems, mutation, migration, selection, and random genetic drift, insofar as these could be deduced a priori from Mendelian inheritance and Darwinian processes. The essence of this theory is contained in Chapters 1 and 2. Its fundamental variables are allele frequencies. But no experimental methods of general utility were available to detect allelic differences between organisms present in natural populations. Apart from a handful of special cases, such as the chromosomal inversions in populations of *Drosophila* that could be studied cytologically, there were almost no allele-frequency data to which the theory could be applied.

This situation has been stood on its head by the application of molecular methods to the study of organisms from natural populations. In Chapter 1 we ingested an alphabet soup of acronyms for different types of DNA polymorphisms: AFLP, RAPD, RFLP, SCAR, SNP, SSCP, and VNTR, among others. The abundance of molecular polymorphisms within species and differences among species offer unprecedented opportunities for inferring mechanisms of evolutionary change and for testing evolutionary hypotheses. Just as some surnames imply a lot about their bearers' ethnic heritage, so do nucleotide and amino acid sequences contain information about their evolutionary heritage. The problem is to be able to interpret that information. Which brings us back to population genetics theory, both classical and modern, because the theory provides the framework upon which any inferences must be based. Nevertheless the focus of population genetics has changed absolutely—from inquiring what deductions can be made about the evolutionary process from the abstract principles of Mendelian inheritance and Darwinian selection, to inquiring what inferences can be made about the evolutionary process from the analysis of sequences of real genes sampled from actual evolving populations.

MOLECULAR POLYMORPHISMS

In Chapters 1 and 2 we were mainly concerned with the analysis of allele frequencies of polymorphic genes. In this chapter we turn to the analysis of nucleotide and amino acid sequences. The data typically consist of a sample of aligned sequences. By a **sequence alignment** we mean that each sequence is so arranged relative to the others, adding gaps where necessary, that each site in the aligned sequences corresponds to a single site in a common ancestral molecule from which the sequences evolved. The sequences may derive from individuals within a single species or from individuals representing two or more species. In this section we focus on samples of sequences from a single species.

The Information Content of Molecular Sequences

Some of the key concepts may be illustrated with an example. The data in Figure 3.1A comprise 500 bp (base pairs) of the coding sequence from five naturally occurring alleles of the *Rh3* (rhodopsin 3) gene of *Drosophila simulans,* extracted from a much larger dataset (Ayala et al. 1993). Several types of nucleotide sites may be distinguished.

- A **segregating site** is one that is polymorphic in the sample. In this case only the 16 segregating sites are listed. They are numbered consecutively, but in fact they are scattered throughout the whole sequence, separated by distances of 2–104 bp, with an average spacing of 22 bp. The sample also contains 484 sites that are monomorphic, each of which is a nonsegregating site. The proportion of segregating sites, typically denoted S, in this sample is therefore $S = 16/500 = 0.0320$.
- A **pairwise difference** between any two sequences is a site at which the sequences differ. The proportion or pairwise differences in a sample is obtained by comparing the sequences in all possible pairs and averaging the number of differences. In the *Rh3* example there are 5 sequences that can be paired in 10 ways. More generally, among n sequences there are $n(n-1)/2$ possible pairwise comparisons. The number of pairwise differ-

(A)

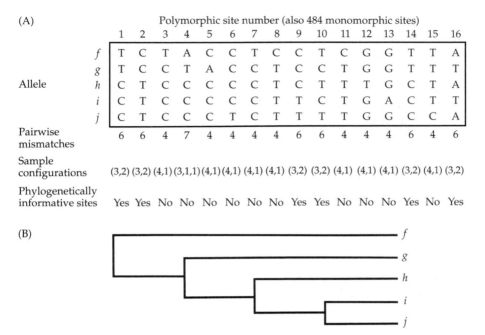

Figure 3.1 (A) DNA polymorphisms among alleles *a–j* of the *Rh3* (rhodopsin 3) gene of *Drosophila simulans*. Each sequence consists of 500 bp of coding sequence. Only polymorphic sites are shown. (Data from Ayala et al. 1993.) (B) Gene tree of the *Rh3* sequences.

ences at each polymorphic site is listed across the bottom of the table. For example, site 1 has 2 T's and 3 C's, and hence $2 \times 3 = 6$ mismatches in pairwise comparisons. The proportion of pairwise mismatches, usually denoted π, is obtained by summing the mismatches across the sample and dividing by the total number of pairwise comparisons. In this case $\pi = (4 \times 9 + 6 \times 6 + 7 \times 1)/(16 \times 10 + 484 \times 10) = 0.0158$, where the denominator includes the 484 monomorphic sites, each of which contributes 0 mismatches.

- The **sample configuration** of a site is the set of numbers giving, in decreasing order, the count of each different kind of element present at a particular site in a sample. Site 1 in the *Rh3* data has the configuration (3, 2, 0, 0), but the 0s are normally omitted and this is written as (3, 2). The symbol (3, 2) means that the sample site includes 3 with the majority nucleotide (in this case C) and 2 with a different nucleotide (in this case T). Site 2 also has the sample configuration (3, 2), although in this case the majority and minority nucleotides are reversed. This means that the sample configurations are indifferent to the identity of the nucleotides at a site, but depend only on the relative numbers. When ties occur both numbers are listed. For example, site 4 has the configuration (3, 1, 1), where each 1 represents a **singleton** occurring only once at the site. In this case the singletons happen to be A and T, but the sample configuration would be (3, 1, 1) regardless. The 484 monomorphic sites all have the sample configuration (5), but this would normally be written (5, 0) to emphasize that the sites are monomorphic.

- A sample of aligned sequences also contains sites that provide information about the **genealogy,** or ancestral relationships, among the sequences. A polymorphic nucleotide site is said to be **phylogenetically informative** if at least one minority nucleotide is not a singleton. Such sites allow the sequences to be split into two groups, each containing two or more members, in which the members of each group are more similar to each other than they are to members of any other group. For example, site 1 in the *Rh3* data is phylogentically informative because the (3, 2) configuration splits the sample into 3 with C at the site and 2 with T at the site. The implication is that, at an earlier time in evolutionary history, the site may have been monomorphic for either C or T, and a nucleotide substitution created a second lineage with the site occupied by the alternative nucleotide. This inference is justified as long as:
  - Each type of nucleotide substitution at a site can occur only once, and
  - There is no reverse mutation that can restore the original nucleotide.

An ancestral history, or **gene tree,** of the *Rh3* samples is shown in Figure 3.1B. In the diagram, time proceeds from left to right. Starting at the far left when all the sequences are assumed to be identical, first the lineage leading to allele *f* split off. Then the lineage leading to *g* split off, followed by *h*, then came the split between *i* and *j*. Stated in another way, the gene tree means that alleles *i* and *j* are the most closely related, with *h* a little more distant, *g* more distant still, and *f* the most divergent. Methods for constructing such trees from sequence data are discussed later in this chapter.

The Coalescent

Although segregating sites, pairwise mismatches, and the gene tree may seem to be independent types of information, they can all be related through the elegant concept of the coalescent. The concept is illustrated in Figure 3.2 for a sample of $n = 6$ alleles whose functional identities are masked by the Greek symbols. As one proceeds back in time the lineages come together, or undergo **coalescence,** where the alleles most recently shared a common ancestor. The coalescence times are indicated by the dashed lines. The symbol T_i represents the random coalescence time at which the lineages of two of *i* alleles in the population coalesce into $i - 1$ lineages. Ultimately, all of the gene genealogies

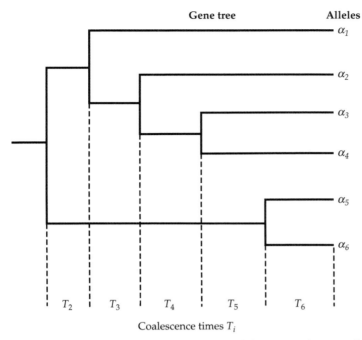

Gene tree   **Alleles**

Figure 3.2 Hypothetical gene tree of six alleles α_1–α_6 showing the coalescence times T_i.

coalesce into one, which is the common ancestor of all of the alleles in the sample of n. The total time back to the common ancestor is given by the sum of the coalescence times T_i. The analysis of coalescence times has become an important part of theoretical population genetics because powerful inferences can be made (for examples, see Hudson 1990; Donnelly and Tavare 1995; Hudson and Kaplan 1995a; Slatkin 1996; Fu and Li 1997; Perlitz and Stephan 1997). Here we give only the briefest introduction to the theory.

Consider i alleles present at some time in a population of size N. What is the probability that there is a coalescence in the immediately preceding generation? This must equal 1 minus the probability that there is no coalescence in the immediately preceding generation, a probability that is quite straightforward to calculate. Figure 2.3 shows that the probability of two alleles in the present generation having distinct ancestors in the previous generation is $1 - 1/(2N) = (2N - 1)/(2N)$. It follows that the probability of a third allele having an ancestor distinct from the first two is $(2N - 2)/(2N) = 1 - 2/(2N)$, because once the distinct ancestors of the first two alleles are chosen, there are only $2N - 2$ distinct ancestors left to choose from. Similarly, the probability that a fourth allele has an ancestor distinct from the first three is $(2N - 3)/(2N) =$

$1 - 3/(2N)$, and so forth. Since these events are independent, the overall probability that i distinct alleles present in any generation have i distinct ancestors in the previous generation is

$$\left(1 - \frac{1}{2N}\right)\left(1 - \frac{2}{2N}\right)\cdots\left(1 - \frac{i-1}{2N}\right) \approx 1 - \frac{1}{2N}(1 + 2 + \cdots + i - 1) = 1 - \frac{i(i-1)}{4N} \quad (3.1)$$

where the approximation assumes that $1/N^2$ is small compared with $1/N$. The simplification on the far right comes from the fact that the sum of the first $i - 1$ integers equals $i(i - 1)/2$.

Since the right-hand side of Equation 3.1 is the probability of the *absence* of a coalescence, the probability of the *presence* of a coalescence, C, is

$$C = 1 - \left[1 - \frac{i(i-1)}{4N}\right] = \frac{i(i-1)}{4N} \quad (3.2)$$

Therefore, for i alleles, the probability of no coalescence for the first $t - 1$ generations followed by coalescence in the tth generation is $(1 - C)^{t-1}C$, and the mean of this geometric distribution of coalescence times is $\bar{T}_i$ where

$$\bar{T}_i = \sum_{t=1}^{\infty} t(1-C)^{t-1}C = \frac{1}{C} = \frac{4N}{i(i-1)} \quad (3.3)$$

Next we discuss an important implication of Equation 3.3 that serves to emphasize the power of coalescence theory.

Nucleotide Polymorphism

In Figure 3.2, note that the interval marked by each T_i has i coexisting lineages. This means that the total time encompassed by all the branches of the gene tree is given by the sum of $i \times T_i$, which has the mean value

$$\sum_{i=2}^{n} i\bar{T}_i = \sum_{i=2}^{n} i\frac{4N}{i(i-1)} = 4N\left(1 + \frac{1}{2} + \frac{1}{3} + \cdots + \frac{1}{n-1}\right) \quad (3.4)$$

In an infinite-alleles mutation model, each new mutation in the branches of the gene tree results in a distinct allele in the sample. This is a plausible assumption for DNA sequences, where a mutation at a particular nucleotide site yields a single nucleotide polymorphism in the sample. If the mutations occur uniformly in time at a rate μ per nucleotide site per generation, then the expected proportion of segregating sites in the sample, $E(S)$, must equal the mutation rate per nucleotide site times the total length of all the branches in the tree, or

$$E(S) = \mu \sum_{i=2}^{n} i\bar{T_i} = 4N\mu\left(1 + \frac{1}{2} + \frac{1}{3} + \cdots + \frac{1}{n-1}\right) = a_1\theta \tag{3.5}$$

where $\theta = 4N\mu$ and

$$a_1 = \left(1 + \frac{1}{2} + \frac{1}{3} + \cdots + \frac{1}{n-1}\right) \qquad a_2 = \left(1 + \frac{1}{2^2} + \frac{1}{3^2} + \cdots + \frac{1}{(n-1)^2}\right) \tag{3.6}$$

(We will make use of a_2 in a moment.) Equation 3.5 implies that θ can be estimated as $\langle\theta\rangle = S/a_1$, which in this context is often called the **nucleotide polymorphism**. One implicit assumption is that the nucleotide sites are completely linked (no recombination). Watterson (1975) has also shown that, with no recombination, the variance of $S/a_1 = \langle\theta\rangle$ is given by

$$\mathrm{Var}\left(\frac{S}{a_1}\right) = \mathrm{Var}\langle\theta\rangle = \frac{\langle\theta\rangle}{ka_1} + \frac{a_2\langle\theta\rangle^2}{a_1^2} \tag{3.7}$$

where k is the number of nucleotides in each aligned sequence, and a_2 is defined in Equation 3.6. Applying these formulas to the *Rh3* data in Figure 3.1, we have $S = 0.0320$, $k = 500$, $a_1 = 2.0833$, and $a_2 = 1.4236$. Hence $\langle\theta\rangle = 0.01536$ and $\mathrm{Var}\langle\theta\rangle = 9.2131 \times 10^{-5}$, which yields a standard error of $\langle\theta\rangle = 0.009598$.

Nucleotide Diversity

The coalescences can also be considered in pairs to derive expressions for the proportion of pairwise differences π as a function of θ. Assuming complete linkage, Tajima (1983) showed that the expected proportion of pairwise differences per nucleotide site, $E(\pi)$, satisfies

$$E(\pi) = \theta \tag{3.8}$$

so that $\langle\theta\rangle = \pi$ is also an estimate of θ. In this context π is often called the **nucleotide diversity**. The variance of the estimate $\pi = \langle\theta\rangle$ in Equation 3.8 is

$$\text{Var}(\pi) = \text{Var}\langle\theta\rangle = \frac{b_1\langle\theta\rangle}{k} + b_2\langle\theta\rangle^2 \tag{3.9}$$

where

$$b_1 = \frac{n+1}{3(n-1)} \qquad b_2 = \frac{2(n^2+n+3)}{9n(n-1)} \tag{3.10}$$

Applying these data to the pairwise differences in the *Rh3* data, we have $n = 5$, $k = 500$, $\pi = 0.01580$, $b_1 = 0.5$, and $b_2 = 0.3667$, hence $\langle\theta\rangle = 0.01580$ and $\text{Var}\langle\theta\rangle = 1.0733 \times 10^{-4}$, which yields a standard error of $\langle\theta\rangle = 0.01036$.

In the *Rh3* data there is very good agreement between the estimate of θ based on nucleotide polymorphism S, from which $\langle\theta\rangle = S/a_1 = 0.01536$, and that based on nucleotide diversity π, from which $\langle\theta\rangle = \pi = 0.01580$. But such good agreement is expected only when the assumptions of the model are satisfied. These include the stipulation that the sample be from a population that is in equilibrium between mutation and random genetic drift, and also that the polymorphisms be selectively neutral.

Tajima's D Statistic

Tajima (1989) proposed that the difference between π and S/a_1 could be used as a test of selective neutrality and other assumptions of the model. The rationale is that nucleotide polymorphism and nucleotide diversity differ primarily because the former is indifferent to the relative frequencies of the polymorphic nucleotides at a site. They lead to consistent estimates for θ anyway, unless some evolutionary process causes a discrepancy. The major discrepancies occur when:

- The frequencies of polymorphic variants are too nearly equal. This pattern increases the proportion of pairwise differences over its neutral expectation, hence $\pi - S/a_1$ is positive. The finding typically suggests either some type of balancing selection, in which heterozygous genotypes are favored, or some type of diversifying selection, in which genotypes carrying the less common alleles are favored.
- The frequencies of the polymorphic variants are too unequal, with an excess of the most common type and a deficiency of the less common types. This pattern results in a decrease in the proportion of pairwise differences, so $\pi - S/a_1$ is negative. Typical reasons for excessively unequal frequencies are:
  - Selection against genotypes carrying the less frequent alleles.
  - A recent population bottleneck eliminating less frequent alleles, and insufficient time since the bottleneck to restore the equilibrium between mutation and random genetic drift.

Tajima's (1989) test is based in a statistic now called **Tajima's D,** which is calculated as:

$$D = \frac{\pi - \dfrac{S}{a_1}}{\sqrt{c_1 S + c_2 S\left(S - \dfrac{1}{k}\right)}} \tag{3.11}$$

where

$$c_1 = \frac{b_1}{a_1} - \frac{1}{a_1^2} \qquad c_2 = \left(\frac{1}{a_1^2 + a_2}\right)\left(b_2 - \frac{n+2}{a_1 n} + \frac{a_2}{a_1^2}\right) \tag{3.12}$$

The denominator in Equation 3.11 is the estimated standard error of $\pi - S/a_1$, hence under the null hypothesis of equilibrium and selective neutrality, the mean and variance of Tajima's D are approximately 0 and 1. Tajima (1989) gives tables to test the significance of a particular value of D, but computer simulations may be more accurate (Fu and Li 1993). For the *Rh3* data, $c_1 = 0.0096$ and $c_2 = 0.003933$. The observed $D = 0.0250$, which is very close to its expected value. If the sample configurations of the 16 polymorphic sites were all (4, 1), then $D = -0.1452$; whereas if the sample configurations were all (2, 1, 1, 1), then $D = 0.7621$. None of these values is significantly different from 0 owing to the small sample size, but they serve to demonstrate how D is affected by sample configurations that are either too unequal (D negative) or too equal (D positive).

PATTERNS OF CHANGE IN NUCLEOTIDE AND AMINO ACID SEQUENCES

The rate of molecular evolution differs from one gene to the next and even from one part of a gene to another. An important principle is that, for neutral mutations, the rate of nucleotide substitution equals the neutral mutation rate (Kimura 1968). To understand why, suppose that at any particular nucleotide site the probability that a new neutral mutation occurs is μ per year. (The units are per year, rather than per generation, because in estimating rates of molecular evolution the absolute time is usually known with greater precision than the number of generations.) With this mutation rate, there will be $2N \times \mu$ such mutations at the site per year in a diploid population of size N. We already know from Equation 2.43 that the probability of a neutral mutation eventually becoming fixed is $1/(2N)$. Hence, in any year, the probability that a nucleotide site undergoes a new neutral mutation that is destined to become fixed is

$$\text{Rate of neutral nucleotide substitutions} = 2N\mu \times \frac{1}{2N} = \mu \tag{3.13}$$

To say the same thing in a slightly different way, the average time between successive fixations of neutral mutations equals $1/\mu$ years. In the next section we shall see that the maximum rate of molecular evolution observed in mammals is approximately 4 nucleotide substitutions per nucleotide site per 10^9 years. Assuming that the maximum rate is observed in a DNA sequence because all nucleotide mutations are selectively neutral, then the rate of mutation would be approximately $1/(4 \times 10^9) = 2.5 \times 10^{-10}$ per nucleotide site per year.

Synonymous and Nonsynonymous Substitutions

The most rapidly evolving DNA sequences include **pseudogenes,** which are homologous to known genes but which have undergone one or more mutations eliminating their ability to be expressed. Pseudogenes are thought to be completely nonfunctional relics of mutational inactivation. This view is supported by their rapid rate of nucleotide substitution, which is shown in comparison with other types of sequences in Figure 3.3.

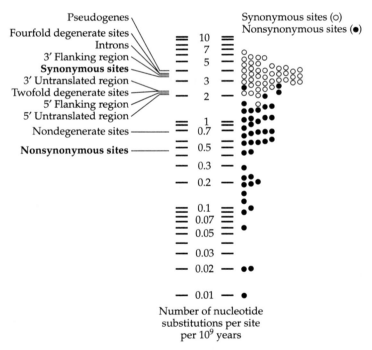

Figure 3.3 Rates of nucleotide substitution in mammalian genes. On the left are the average rates for different types of sequences. On the right are rates for synonymous and nonsynonymous sites in a sample of 43 genes. (Data from Li 1997.)

Note that the scale in Figure 3.3 is logarithmic, which means that there is a 1000-fold difference in substitution rate from the bottom to the top. Other gene regions are defined as in Figure 1.7A (page 12), and the different kinds of degenerate sites (Li et al. 1985) are defined according to the synonymous codons in the standard genetic code shown in Table 1.1 (page 5). The codons for eight amino acids can have any nucleotide (U, C, A, or G) in their third position, which is therefore known as a **fourfold degenerate site.** The codons for seven amino acids can terminate in either U or C, and those for five amino acids can terminate in either A or G; these third-position sites are called **twofold degenerate sites.** A site is **nondegenerate** if any nucleotide substitution changes the amino acid encoded. Because of the codon degeneracies, nucleotides in a coding sequence can change without affecting the amino acid sequence. These changes are called **synonymous substitutions,** and their rate is usually symbolized as K_s. Figure 3.3 indicates that the average rate of synonymous substitution in mammals is $K_s = 3.4 \times 10^{-9}$ synonymous substitutions per synonymous site per year, where the number of synonymous sites is calculated as the number of fourfold degenerate sites plus 1/3 the number of twofold degenerate sites. The fraction 1/3 is used because, with random mutation at twofold degenerate sites, 1/3 of all nucleotide substitutions are expected to result in a synonymous codon. On the other hand, nucleotide substitutions that do change amino acids are **nonsynonymous substitutions** or **amino acid replacements,** and their rate is usually symbolized as K_a. The overall average rate of nonsynonymous substitutions in mammals is $K_a = 0.46 \times 10^{-9}$ nonsynonymous substitutions per nonsynonymous site per year, where the number of nonsynonymous sites is defined as the number of nondegenerate sites plus 2/3 the number of twofold degenerate sites.

Especially among nonsynonymous substitutions, there is very great variation from one gene to the next. This is shown by the filled circles at the right in Figure 3.3. The large variation in rates is attributed to selective constraints on amino acid replacements that do not operate as strongly on synonymous nucleotide substitutions. A **selective constraint** means the tendency for certain amino acids to be conserved at particular sites because of natural selection for optimal function. Not just any amino acid will serve at a particular position in a protein molecule, because each amino acid must participate in the chemical interactions that fold the molecule into its three-dimensional shape and give the molecule its specificity and ability to function. The need for proper chemical interactions and folding constrains the acceptable amino acids that can occupy each site. Although some amino acid replacements may be functionally equivalent or nearly equivalent, many more are expected to impair protein function to such an extent that they reduce the fitness of the organisms that contain them. From comparisons of synonymous and nonsynonymoous substitutions in hominid lineages, Eyre-Walker and

Keightley (1999) have estimated that an average of approximately 45% of all amino acid replacements that have occurred in the human, chimpanzee, and gorilla lineages have been eliminated by natural selection. This average is substantially smaller than the approximately 75% elimination estimated in the primate, artiodactyl, and rodent lineages. The difference is attributed to a smaller effective population number in hominids, which allows more slightly deleterious mutations to become fixed. Eyre-Walker and Keightley (1999) also estimate that each diploid human genome suffers 1.6–3.1 new deleterious mutations per generation (0.064–0.103 per year). Most of these new mutations are destined to be lost through differential survival and reproduction. But the sheer number of deleterious mutations is too great for them to be eliminated independently (additive or multiplicative fitnesses). More likely they are eliminated in bunches through synergistic interactions on fitness (Crow 1999).

Codon Usage Bias

The synonymous substitutions in Figure 3.3 (open circles) occupy a relatively narrow range of rates, but some synonymous substitutions are also constrained. One type of constraint occurs through secondary structures that form in some RNA molecules due to internal base pairing (Parsch et al. 1998) A second, more extensively studied type of constraint occurs through synonymous codon preferences, which are correlated with the relative abundance of tRNA molecules used to translate the codons. In bacteria and yeast, highly expressed proteins tend to use synonymous codons for tRNA molecules that are abundant in the cell, whereas proteins produced in small amounts do not show such codon usage bias (Ikemura 1985; Eyre-Walker 1996). Natural selection also acts to optimize codon usage in such organisms as *Drosophila* (Akashi 1993, 1995), but in warm-blooded vertebrates codon usage bias is confounded with large-scale differences in base composition (G + C content) and patterns of nucleotide substitution across different regions of the genome (Bernardi 1995).

Nonrandomness in synonymous codon usage can be measured in a variety of ways. The most commonly encountered measures are the following:

- The **effective number of codons (ENC)** is a measure of departure from equal codon usage that is independent of gene length, amino acid composition, and any reference set of genes (Wright 1990). For each amino acid, the ENC is the number which, if used equally frequently, would yield the same sum of squares of the actual codon frequencies used for that amino acid. The ENC for the entire gene is the sum of the ENCs for each amino acid. A low ENC corresponds to high codon usage bias, a high ENC to low codon usage bias. The minimum value of 20 occurs when one codon is used exclusively for each amino acid, and the maximum is 61 when synonymous codons are used equally. The typical range of ENC is 25–55.

- The **scaled chi-square** (χ^2/L) is another measure of deviation from equal codon usage (Shields et al. 1988). For each synonymous codon group, a χ^2 value is calculated according to Equation 1.5. The scaled value is calculated by summing these values for each synonymous codon group and dividing by the total number of codons (L) in the gene. The scaled chi-square theoretically ranges from 0–1, and the observed range is almost as wide, but most genes are in the range 0.1–0.6.

- The **codon adaptation index (CAI)** estimates the extent to which codon usage is biased toward codons used in a set of highly expressed reference genes in the same species (Sharp and Li 1987). Each codon is assigned a relative "adaptiveness" value according to its frequency of use in the reference genes. The CAI for any gene is then calculated as the geometric mean of the relative adaptiveness of its codons, divided by the geometric mean of the relative adaptiveness if each codon used were optimal. (The geometric mean is the Lth root of the product of L numbers, where in this case L is the number of codons in the gene.) A CAI near 1 means high codon usage bias; near 0 means that the gene has an extensive concentration of otherwise rarely used codons. In yeast the CAI ranges from about 0.1–0.9 and in *E. coli* from about 0.2–0.8, with highly expressed genes having the greater CAI.

Selection for Optimal Codons and Amino Acids

The pattern of codon usage bias in a gene exemplifies the interplay between mutation, RGD, and weak selection. Each codon in a gene represents a sort of "replicate experiment" that, at any point in time, may be fixed for an optimal codon, fixed for a nonoptimal codon, or be polymorphic. To analyze this situation quantitatively, consider a coding sequence containing k_s synonymous sites evolving in a haploid organism such as yeast or *E. coli*. In any generation, any of the sites can undergo mutation with probability μ per site. Let us assume that each site evolves independently, which means that the rate of recombination between sites is at least on the order of the mutation rate per site. Let us assume furthermore that, once a site mutates, its evolutionary fate of fixation or loss is resolved before the same site mutates again. The properties of independence with at most two nucleotides at any polymorphic site define the **Poisson random field (PRF)** model of molecular evolution (Sawyer and Hartl 1992). Alternative models that allow multiple nucleotides at a site or linkage between sites are discussed in Ewens (1972) and Watterson (1985). In the PRF model, each new mutation initiates an independent process of RGD with selection that can be modeled as a diffusion process of the type discussed in Chapter 2.

Suppose now that n sequences are sampled from the population. Using the PRF model it can be shown that the expected number of polymorphic sites with exactly r mutant and $n - r$ nonmutant nucleotides ($r = 1, 2, 3, ..., n - 1$) has a Poisson distribution with mean

$$\lambda_r = 2N\mu k_s \int_0^1 \frac{1-e^{-2Ns(1-p)}}{1-e^{-2Ns}} \frac{1}{p(1-p)} \times \frac{n!}{r!(n-r)!} p^r (1-p)^{n-r} dp \quad \text{if } s \neq 0 \quad (3.14)$$

$$\lambda_r = 2N\mu k_s \int_0^1 \frac{1}{p} \times \frac{n!}{r!(n-r)!} p^r (1-p)^{n-r} dp \quad \text{if } s = 0 \quad (3.15)$$

In these equations, the factor following the times sign is the binomial sampling distribution, and the factor preceding the times sign is the limiting probability distribution of mutant nucleotides. For a **Poisson distribution** with mean λ, the probability that a particular sample leads to a value $i = 0$, 1, 2, ... is given by

$$\Pr\{\text{observed number} = i\} = e^{-\lambda} \frac{\lambda^i}{i!} \quad (3.16)$$

Furthermore, under the PRF model, the expected numbers given in Equations 3.14 and 3.15 are independent for $1 \leq r \leq n - 1$. Equations 3.14 and 3.15 therefore yield the expected numbers of each of the sample configurations of polymorphic sites. Since a polymorphic site with the sample configuration $(n - r, r)$, where $n - r \geq r$, consists either of r mutant nucleotides and $n - r$ nonmutant nucleotides, or of r nonmutant nucleotides and $n - r$ mutant nucleotides, the overall expected number of sites with the sample configuration $(n - r, r)$ is given by $\lambda_r + \lambda_{n-r}$ from Equations 3.14 or 3.15.

An application of the PRF sampling theory to actual data is shown in Figure 3.4, where the sample consists of 14 sequences of the *E. coli* gene *gnd* encoding 6-phosphogluconate dehydrogenase (Hartl et al. 1994). The expected numbers were obtained by choosing values of $N\mu$ and Ns in Equations 3.14 and 3.15 so as to maximize the probability of obtaining the observed data under these values. The sample configurations for synonymous polymorphisms (dots) show a significant departure from the neutral expectation, but they do fit a curve with $Ns = -1.34$, where s is the selection coefficient against nonoptimal synonymous codons. The same approach can be used to analyze the sample configurations of polymorphic amino acids (squares), replacing k_s with the number of nonsynonymous nucleotide sites

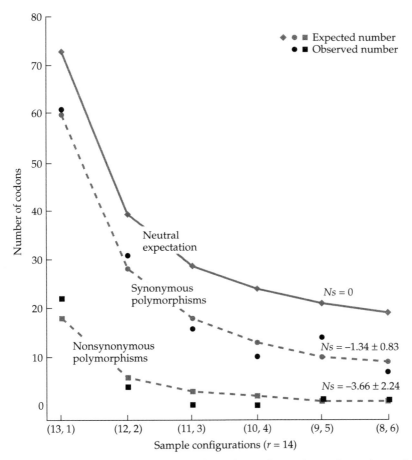

Figure 3.4 Observed and expected numbers of sample configurations of synonymous and nonsynonymous polymorphisms in the *gnd* gene for 6-phosphogluconate dehydrogenase in *E. coli*. (Data from Hartl et al. 1994.)

k_a. The amino acid polymorphisms fit a model with $Ns = -3.66$, which implies that there is about a 2.7-fold greater selection against deleterious amino acid replacements than against nonoptimal codons in this gene. If we take the analysis one step further, the estimate of $N\mu$ for these data is 0.092, but in *E. coli* μ has been estimated directly as $\mu = 5 \times 10^{-10}$ (Drake 1991). Hence, we can estimate the effective population number N as $N = 0.092/\mu = 1.8 \times 10^8$. If we use this value of N and the previous estimates of Ns, the average selection coefficient against a nonoptimal synonymous codon is 7.4×10^{-9} and the average selection coefficient against a deleterious polymorphic amino acid replacement is 2.0×10^{-8}. These selection coefficients are very small, but in a population so large the selection coefficient against a deleterious polymorphism must be small or else the mutation would not be polymorphic. For the sake of convenience we have assumed that the selection rate is the same for all new mutations, or at least for all new mutations that have any chance of becoming polymorphic in the population. Our estimates of s are therefore an average taken across different mutable sites. More realistic models could incorporate a probability distribution for the selection coefficient, but this is unnecessary for present purposes.

Selective Sweeps versus Background Selection

The level of nucleotide diversity π across the genome of *Drosophila* decreases with the rate of genetic recombination (Begun and Aquadro 1992). This pattern is illustrated in Figure 3.5. Two diametrically opposed hypotheses have been put forward to explain this finding. One attributes it to beneficial mutations, and the other to deleterious mutations. The process in which a strongly beneficial mutation rapidly becomes fixed in a population is called a **selective sweep.** During a selective sweep of a beneficial mutation, any neutral alleles that are sufficiently tightly linked in the chromosome will also increase in frequency **(genetic hitchhiking)**. The hitchhiking of linked genes reduces the amount of genetic variation in a chromosomal region flanking the beneficial mutation. If selective sweeps occur frequently enough, regions of low recombination will seldom, if ever, have a chance to regain genetic variation by mutation and RGD before the recovery process is interrupted by the next beneficial mutation (Perlitz and Stephan 1997).

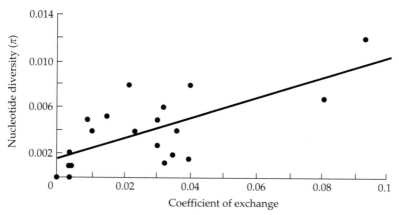

Figure 3.5   Relation between nucleotide diversity and rate of recombination among genes in *D. melanogaster*. (From Begun and Aquadro 1992.)

In the other model, called **background selection,** the low level of polymorphism in regions of tight linkage is attributed to deleterious mutations. Each deleterious mutation that occurs dooms some little region of chromosome in its vicinity to eventual extinction. The lower the rate of recombination, the larger the region of chromosome that is doomed. In effect, each new deleterious mutation reduces by one the number of chromosomes that can contribute to remote future generations. If there is complete linkage (no recombination), as in *Drosophila* chromosome 4, then the whole chromosome is affected; if there is recombination, then only a region around the mutation is affected. In either case, a sufficient density of deleterious mutations will reduce the number of surviving lineages to such an extent that the degree of polymorphism will be smaller than expected, given the actual population size, and the tighter the linkage the greater the disparity (Charlesworth et al. 1995; Hudson and Kaplan 1995b). In effect, background selection reduces the effective population size for regions of tight linkage, thereby reducing the nucleotide diversity.

The main effect of a selective sweep is that a small region around the beneficial mutation will be overrepresented in the population. This results in frequencies of polymorphic variants that are too unequal, owing to the overrepresentation of alleles that profited from the hitchhiking, and Tajima's D will be negative. With background selection, on the other hand, although there is a reduction in the level of polymorphism, there is no skewing of the distribution of frequencies. For all practical purposes, the deleterious allele merely causes one chromosome to drop out of the population, much as if it were to go extinct by chance. In this case, Tajima's D should be close to its expected value of 0. Analysis of several *Drosophila* datasets has failed to detect the significantly negative D that would be expected from recurrent selective sweeps (Braverman et al. 1995).

One of the hallmarks of persistent strong background selection is a reduction in codon usage bias due to small effective population size (Kliman and Hey 1994; Munte et al. 1997). This effect is exemplified by the *Drosophila* genes numbered 1 through 4 in Figure 3.6, which are located deep in the centromeric heterochromatin of chromosome 2 (gene 1) or that of the X chromosome (2–4), where recombination is severely restricted. Genes 5–17 are located somewhat farther out along the X chromosome, but still quite near the base. The degree of codon usage bias shows an extremely sharp increase as the gene locations move outward from the centromeric heterochromatin, suggesting that even a small amount of recombination is sufficient to offset the effects of background selection. In *D. melanogaster* there appears to have been at least one selective sweep accompanying the evolution of a completely new gene for a sperm axonemal dynein in the region flanked by genes 10 and 11, since the nucleotide diversity (π) in this region is at least four standard deviations lower than that of gene 17, and the codon usage bias across the region

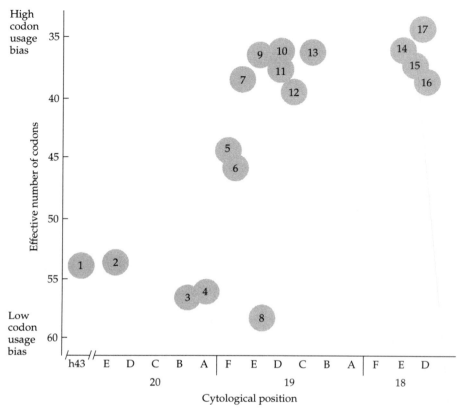

Figure 3.6 Relation between codon usage bias and chromosomal position for genes 2–17 near the base of the X chromosome in *D. melanogaster*. Gene 1 is deep in the centromeric heterochromatin of chromosome 2, where the rate of recombination is very low. Cytological region 20 in the X chromosome is closest to the centromeric heterochromatin. The genes are: 1, *rl*; 2, *su(f)*; 3, *S6kII*; 4, *fog*; 5, *sol*; 6, *slgA*; 7, *dod*; 8, *shakB*; 9, *run*; 10, *AnnX*; 11, *Cdic*; 12, *Pbprp2*; 13, *Pp4–19C*; 14, *Mer*; 15, *Cdc42*; 16, *Bap*; and 17, *Zw*. (From Nurminsky and Hartl 1999.)

argues against strong background selection (Nurminsky et al. 1998; Nurminsky and Hartl 1999; Charlesworth and Charlesworth 1999).

POLYMORPHISM AND DIVERGENCE

In this section we examine the relation between molecular polymorphism within species and genetic divergence between species. This is an important issue because the essence of Darwinism is sometimes asserted to be that polymorphism within species is transformed into divergence between species.

While true as far as it goes, the assertion is all too easily misinterpreted. It does not assert, for example, that the totality of genetic variation within species becomes transformed into genetic differences between species. Nor does it claim that genetic divergence between species must mirror the types and frequencies of genetic polymorphisms within species. For Darwin, speciation was above all a process of adaptation by means of natural selection. But it is only in the absence of selection that there is a smooth, unbiased transformation of polymorphism into divergence.

The Hudson-Kreitman-Aguadé (HKA) Test

How fast does divergence accumulate between reproductively isolated species? We know from Equation 3.8 that, with no selection, the number of pairwise differences between two sequences chosen at random from within a species depends on $\theta = 4N\mu$, so this is the minimum divergence we should expect between sequences from different species. We also know from Equation 3.13 that the rate of evolution with no selection equals the neutral mutation rate, so this must be the rate at which divergence between species accumulates over and above the amount present within a species. Hence, if k is the number of nucleotide sites in a sequence and μ is the neutral mutation rate per nucleotide site, then the expected number of pairwise differences $E(D)$ between a pair of homologous sequences, one chosen from each of two different species, should be

$$E(D) = 2\mu k t_{div} + k\theta = \frac{4N\mu k t_{div}}{2N} + k\theta = k\theta(T+1) \qquad (3.17)$$

where t_{div} is the divergence time in generations and N is the effective population number in each species. The factor of 2 after the first equal sign comes from the fact that, for each species, t_{div} is the number of generations back to the common ancestor, and so the total time elapsed in both branches of the phylogenetic tree is $2t_{div}$. The simplification on the right comes from defining the divergence time T in units of $2N$ generations, so that $T = t_{div}/2N$. As noted below it is not necessary to assume equality in population size. The variance in the number of pairwise differences is also known (Li 1977; Gillespie and Langley 1979), namely,

$$\text{Var}(D) = k\theta(T+1) + k^2\theta^2 \qquad (3.18)$$

Hudson, Kreitman, and Aguadé (1987) devised a statistical test for the selective neutrality of within-species polymorphism and between-species divergence based on these equations. The rationale of the test is that, with no selection, the equilibrium polymorphism depends on θ but the divergence depends on θ and T. Both parameters can therefore be estimated by combining data on polymorphism and divergence, and these estimates can be used in a test for goodness of fit to the hypothesis of no selection. The data for the test consist of a sample of each L genes (or gene regions) from each of two species A and B, with $L \geq 2$. The test statistic for the **HKA test** is

$$X^2 = \sum_{i,j} \frac{\left[S_{ij} - E\left(S_{ij}\right)\right]^2}{\mathrm{Var}\left(S_{ij}\right)} + \sum_i \frac{\left[D_i - E(D_i)\right]^2}{\mathrm{Var}(D_i)} \tag{3.19}$$

where the first summation is for polymorphisms over all genes or gene regions i ($i = 1, 2, ..., L$) in both species $j = A$ or B, and the second summation is for divergence over all genes or gene regions. In this equation, S_{ij} is the number of polymorphic sites for gene i in species j ($j = A$ or B), and D_i is the number of pairwise differences between sequences from the two species. If k_{ij} is the number of nucleotide sites in gene i in species j, then from Equations 3.5 and 3.7,

$$E(S_{ij}) = a_1 k_{ij}\hat{\theta}_i \qquad \mathrm{Var}(S_{ij}) = a_1 k_{ij}\hat{\theta}_i + a_2(k_{ij}\hat{\theta}_i)^2 \tag{3.20}$$

where a_1 and a_2 are determined by the sample sizes according to Equations 3.6 and $\hat{\theta}_i$ is the estimate of θ_i obtained by combining all the data. Equations 3.20 explicitly include the number of nucleotide sites because S_{ij} is the absolute number of polymorphisms, whereas in Equations 3.5 and 3.7, S is the polymorphism per nucleotide site ($S = S_{ij}/k_{ij}$). For the divergence terms, $E(D_i)$ and $\mathrm{Var}(D_i)$ are given by Equations 3.17 and 3.18 for the number of nucleotide sites compared (k_i), the estimated $\hat{\theta}_i$, and the estimated time $\hat{T}$. In general, the effective population sizes need not be assumed to be equal, in which case replace $T + 1$ Equations 3.17 and 3.18 with $T + (1 + f)/2$, where the effective population sizes of A and B at the time of sampling are $2N$ and $2Nf$. Methods for estimating the L parameters $\hat{\theta}_i$ along with $\hat{T}$ and $\hat{f}$ are discussed in detail in Hudson et al. (1987) and Li (1997). Furthermore, Hudson et al. (1987) show that for sufficiently large samples of each gene, X^2 is approximately chi-square distributed with $2L - 2$ degrees of freedom, so that Figure 1.11 can be used to obtain the significance level of the X^2. The test is conservative in rejecting the hypothesis of no selection, since it assumes complete linkage within each gene and free recombination between genes.

The first application of the HKA test was to two regions ($L = 2$) in the alcohol dehydrogenase gene, *Adh*, of *D. melanogaster* and its sibling species

D. sechellia (Hudson et al. 1987). The gene regions consisted of the 5' flanking sequence (region 1) and the synonymous sites in the coding sequence (region 2). For *D. melanogaster* (species A) the sample size in both regions was $n = 81$ (so that $a_1 = 4.9655$ and $a_2 = 1.6325$) and the number of nucleotides was $k_{11} = 414$ for region 1 and $k_{21} = 79$ for region 2. The observed number of polymorphic sites for regions 1 and 2 were $S_{11} = 9$ and $S_{21} = 8$. In *D. sechellia* only one sequence was examined (so the terms involving S_{12} and S_{22} in Equation 3.19 must be dropped). Among $k_1 = 4052$ nucleotide sites compared in region 1 there were $D_1 = 210$ differences, and among $k_2 = 324$ synonymous sites compared in region 2 there were $D_2 = 18$ differences. Because S_{12} and S_{22} are missing data, not all parameters can be estimated. This ambiguity was handled by assuming equal effective population sizes, setting $f = 1$. The estimates obtained for θ per nucleotide site and the divergence time T were

$$\hat{\theta}_1 = 0.0066 \quad \hat{\theta}_2 = 0.0090 \quad \hat{T} = 6.73$$

Hence

$$
\begin{aligned}
X^2 = {} & \frac{\left(9 - 4.9655 \times 414 \times 0.0066\right)^2}{4.9655 \times 414 \times 0.0066 + 1.6325 \times \left(414 \times 0.0066\right)^2} \\[2mm]
& + \frac{\left(8 - 4.9655 \times 79 \times 0.0090\right)^2}{4.9655 \times 79 \times 0.0090 + 1.6325 \times \left(79 \times 0.0090\right)^2} \\[2mm]
& + \frac{\left[210 - 4052 \times 0.0066 \times \left(6.73 + 1\right)\right]^2}{4052 \times 0.0066 \times \left(6.73 + 1\right) + \left(4052 \times 0.0066\right)^2} \\[2mm]
& + \frac{\left[18 - 324 \times 0.0090 \times \left(6.73 + 1\right)\right]^2}{324 \times 0.0090 \times \left(6.73 + 1\right) + \left(324 \times 0.0090\right)^2} \\[2mm]
= {} & 6.07
\end{aligned}
$$

This X^2 has one degree of freedom (since three parameters were estimated from the data), from which Figure 1.11 indicates that $P \approx 0.016$. The hypothesis of neutrality can therefore be rejected with some confidence. But what does this mean? Among the possibilities:

- Too little polymorphism in the 5' flanking region
- Too much polymorphism at the synonymous sites
- Too little divergence at the synonymous sites
- Too much divergence in the 5' flanking region

Hudson et al. (1987) give arguments why they prefer the second explanation, but they also emphasize that the HKA test is an omnibus test that cannot, by itself, identify which of the possibilities, or which combination of possibilities, is the true underlying cause.

The McDonald-Kreitman Test

One approach of commendable simplicity that also tests neutrality of polymorphism and divergence was proposed by McDonald and Kreitman (1991). The test is outlined in Figure 3.7A, which deals with sequences of the alcohol dehydrogenase gene *Adh*. The raw data consist of a set of aligned protein-coding sequences, *m* from one species (in this case *D. simulans*) and *n* from a closely related species (in this case *D. yakuba*). All variable sites are first classified as either synonymous (if the nucleotide substitution results in a synonymous codon), or else as replacement (if the nucleotide substitution results in an amino acid replacement in the protein). Each variable site is then classified as to whether it is polymorphic in one or both species, and finally monomorphic sites that differ in the species are classified as divergent. This dual classification allows each variable site to be placed into one

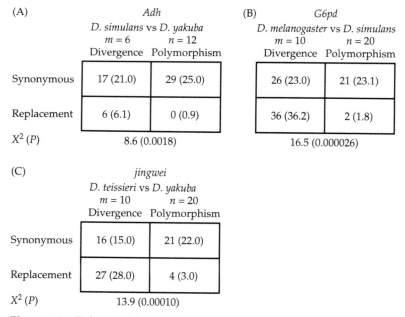

Figure 3.7 Polymorphism–divergence tables for synonymous versus replacement differences for three genes in interspecific comparisons in *Drosophila*. (Data from (A) McDonald and Kreitman 1991; (B) Eanes et al. 1993; (C) Long and Langley 1993.)

of four classes in a 2×2 table, as shown in the figure. The actual counts are the numbers not enclosed in parentheses. (Those in parentheses will be discussed in the next section.)

McDonald and Kreitman realized that if polymorphism within species is faithfully transformed into divergence between species, then the cells in the 2×2 table should be independent. This means that the ratio of divergence to polymorphism should be the same in both synonymous and replacement sites, and that the ratio of synonymous to replacement sites should be the same in the divergent and polymorphic categories. One way to test for independence is through the X^2 statistic calculated from Equation 1.13. These X^2 values and the corresponding probabilities (P) from Figure 1.11 are shown at the bottom. Some of the cells have very few counts but, in fact, the P values obtained from the X^2 are all very close to those obtained by more exact methods. The point of the highly significant P values is that the hypothesis of independence can be decisively rejected. McDonald and Kreitman noted that there seemed to be an excess of divergent over polymorphic replacement sites, relative to what would be expected from the ratio of divergent to polymorphic synonymous sites. If synonymous sites are at most weakly selected, they argued, then the excess of divergent replacement sites must be the result of adaptive evolution incorporating beneficial amino acid replacements into the protein. Darwinism *revanche*!

The other data in Figure 3.7 are for the genes *G6pd* (Eanes et al. 1993), which encodes the metabolic enzyme glucose-6-phosphate dehydrogenase, and *jingwei* (Long and Langley 1993), which is a gene of unknown function that evolved relatively recently in the *D. teissieri/D. yakuba* lineage after co-opting some of the exons of *Adh*. These genes also show the signature of adaptive amino acid replacements in having an excess of replacement divergence over replacement polymorphisms, relative to that seen for synonymous sites. But not all genes show the patterns exhibited in Figure 3.7. For example, analogous 2×2 tables for the *Drosophila* genes *period* (Kliman and Hey 1993), *zeste* and *yolk protein 2* (Hey and Kliman 1993), and *bride of sevenless* (Ayala and Hartl 1993) do not show any significant departure from independence. This suggests that not all genes are undergoing adaptive evolution all the time. The examples in Figure 3.7 are somewhat special, too, because there were a priori reasons to expect adaptive protein evolution. The *Adh* gene is a well known target of selection in *D. melanogaster* (van Delden et al. 1978; David et al. 1986; Kreitman and Hudson 1991; Stam and Laurie 1996), both *Adh* and *G6pd* show a geographical gradient, or **cline,** of allozyme frequencies strongly suggestive of selection (Oakeshott et al. 1982, 1983; Berry and Kreitman 1993), and *jingwei* is a new gene still likely to be acquiring its optimal function (Long and Langley 1993). For additional discussion see Brookfield and Sharp (1994) and Eanes et al. (1996).

Polymorphism and Divergence in a Poisson Random Field

The McDonald-Kreitman test is disarming in its simplicity, yet many parameters affecting the outcome are hidden beneath the surface. These parameters include:

- μ_s, the mutation rate per site at synonymous sites, including monomorphic synonymous sites
- μ_r, the mutation rate per site at replacement (nonsynonymous) sites, including monomorphic replacement sites
- N, the effective population size, assumed here to be the same in both species
- s_s, which is negative, the selection coefficient against nonoptimal synonymous codons
- s_r, the selection coefficient for (if positive) or against (if negative) amino acid replacements
- T, which equals $t_{div}/2N$, the divergence time in units of $2N$ generations

The expected number in each cell in the polymorphism–divergence table can be expressed in terms of these parameters by invoking the Poisson random field (PRF) theory discussed earlier in the context of codon usage bias (Equations 3.14–3.16). As noted, the main assumptions of the theory are that each nucleotide site can be segregating for at most two nucleotides at any given time, and that the sites are independent. Under the PRF model, Sawyer and Hartl (1992) showed that the expected numbers in the polymorphism–divergence table are given by the formulas in Figure 3-8. The U's and γ's are composite parameters defined as follows:

- $U_s = \mu_s \times k_s \times N$, where k_s is the number of synonymous sites, including monomorphic synonymous sites
- $U_r = \mu_r \times k_r \times N$, where k_r is the number of replacement (nonsynonymous) sites, including monomorphic replacement sites

Note that μ_s will in general differ from μ_r even though both are mutation rates per nucleotide site, because the rates include only those mutations that have a chance of becoming polymorphic or divergent in the sample. For example, many amino acid replacements are so deleterious in their effects on fitness that they could not possibly contribute to polymorphism or divergence, and these mutations are not included in μ_r. Based on the very small number of amino acid replacements in the *Adh* data in Figure 3.7A, Hartl and Sawyer (1992) estimated that, at any one time, only approximately 6 of the 256 amino acids in the *Adh* protein are susceptible to a favorable amino acid replacement. The other composite parameters in Figure 3.8 are

- $\gamma_s = N \times s_s$, the value of Ns for synonymous sites
- $\gamma_r = N \times s_r$, the value of Ns for replacement sites

	Divergence	Polymorphism
Synonymous	$2U_s \dfrac{2\gamma_s}{1-e^{-2\gamma_s}}[T+G(m)+G(n)]$	$2U_s \dfrac{2\gamma_s}{1-e^{-2\gamma_s}}[F(m)+F(n)]$
Replacement	$2U_r \dfrac{2\gamma_r}{1-e^{-2\gamma_r}}[T+G(m)+G(n)]$	$2U_r \dfrac{2\gamma_r}{1-e^{-2\gamma_r}}[F(m)+F(n)]$

Figure 3.8 Expected numbers for the entries in polymorphism–divergence tables, based on the Poisson random field model. (Formulas from Sawyer and Hartl 1992.)

The functions G and F in Figure 3.8 are defined as

$$G(m)=\int_0^1 (1-x)^{m-1}\frac{1-e^{-2\gamma x}}{2\gamma x}dx \qquad F(m)=\int_0^1 \frac{1-x^m-(1-x)^m}{1-x}\frac{1-e^{-2\gamma x}}{2\gamma x}dx \quad (3.21)$$

where γ should be replaced with γ_s in the upper two formulas and with γ_r in the lower two. As in Figure 3.7, m is the number of sequences from species A and n is the number of sequences from species B. For any specified values of the parameters, these integrals can be evaluated numerically using any of a number of mathematical analysis programs. (I use Mathematica®.) If one is willing to assume that $\gamma_s = 0$ (no selection for optimal codon usage), then the formulas in the top row in Figure 3.8 should be replaced with

$$2U_s\left(T+\frac{1}{m}+\frac{1}{n}\right) \quad \text{and} \quad 2U_s\left[a_1(m)+a_1(n)\right] \quad \text{where } a_1(m)=\sum_{i=1}^{m-1}\frac{1}{i} \quad (3.22)$$

When the expected numbers are shown in the form of a polymorphism–divergence table as in Figure 3.8, the model appears overspecified, as there are five parameters (U_s, U_r, γ_s, γ_r, and T) but only four observations. This is deceptive, however, since the sample configurations of the synonymous sites allow independent estimation of U_s and γ_s using Equation 3.14 (see Figure 3.4). The sample configurations also allow independent estimation of T (Sawyer and Hartl 1992), but T is also often available from independent datasets.

What intensities of selection are implied by the results in Figure 3.7? From the PRF model the estimates are:

	Adh	G6pd	jingwei
$\gamma_s = N \times s_s$ (synonymous)	−1	0	−0.75
$\gamma_r = N \times s_r$ (replacement)	+10	+20	+6

As seen before in data from *E. coli* (Figure 3.4), two of the genes show relatively weak selection against nonoptimal synonymous codons. This result is consistent with findings of Akashi (1995), who used a different approach. For all three genes the excess of amino acid replacements can be explained by the fixation of beneficial mutations, and in all cases the magnitude of selection, as measured by Ns, is quite large. These estimates should be interpreted as averages, since the PRF model makes the probably unrealistic assumption that all amino acid replacements that are candidates for polymorphism or divergence have the same selective effect. In Figure 3.7, the numbers in parentheses are the expected values from the PRF model with these parameters. It might finally be noted that, while $Ns = +10$ is certainly large enough so that the fate of a polymorphic allele will be determined primarily by selection, the value of s still falls very far below any that could be detected in laboratory experiments. The reason is that the effective population size of these *Drosophila* species is estimated to be on the order of 10^6 (Berry et al. 1991; Hartl and Sawyer 1992; Akashi 1995). If $Ns = +10$, this implies that $s = 0.00001$. Darwin (1859) envisaged adaptive evolution as the gradual accumulation of very slight improvements over eons of geological time. The selection coefficients so far estimated from molecular population genetics support his point of view.

Convergence to Adaptation

A given change in enzyme activity in a metabolic pathway may result in a disproportionately smaller change in relative fitness because of the type of "diminishing returns" relation between enzyme activity and relative fitness illustrated in Figure 3.9. In this example the enzyme is β-galactosidase in *E. coli*, and the relative fitnesses are estimated from the results of experiments analogous to those in Figure 2.8 for strains undergoing competition for uptake and metabolism of the sugar lactose, the substrate of β-galactosidase (Hartl et al. 1985). Although few experiments directly assaying relative fitness as a function of enzyme activity have been carried out, considerations of the rate of flux of metabolites through metabolic pathways (Kacser and Burns 1973) suggests that the qualitative result might hold rather generally. The shape of the curve is a hyperbola, like that of first-order enzyme kinetics, with relative fitness (w) as a function of enzyme activity (a) given by

$$w(a) = C\left(\frac{a}{1+a}\right) \tag{3.23}$$

where C is a scaling constant (in this case $C = 31/30$) determining the value of a at which $w = 1$. The hyperbolic shape of the curve implies that, for cells

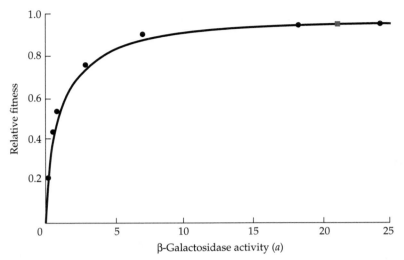

Figure 3.9 Hyperbolic relation between relative fitness of strains of *E. coli* undergoing competition for lactose against relative activity of the β-galacto-sidase enzyme. The gray square represents the laboratory strain K12; the other points are from mutants. (Data from Hartl et al. 1985.)

with enzyme activities near that of wildtype (square), a small change in enzyme activity results in a disproportionately smaller change in fitness. Such a curve was originally invoked by Wright (1934) to explain why most mutations are recessive in their effects on phenotype. In effect, the wildtype level of enzyme activity is high enough to buffer even a twofold decrease in activity. For the curve in Figure 3.9, for example, a 50% decrease in enzyme activity from $a = 30$ to 15 changes the relative fitness by only 3.1% (from $w = 1.000$ to $w = 0.969$).

Different enzymes along a metabolic pathway may differ greatly in how much a given change in activity will changes fitness. For example, in popu-lations of *E. coli* undergoing competition for lactose, the fitness of a mutant strain, relative to the standard *E. coli* K12 strain, is given approximately by

$$\frac{1}{w} = \frac{0.004}{Z} + \frac{0.130}{Y} + 0.866$$

where Z is the activity of the β-galactosidase and Y that of the β-galactoside permease (for uptake of the sugar), both measured relative to the activities in the K12 strain (Dykhuizen et al. 1987). The differing effects of perturbations of activity can be appreciated by considering two mutations, one of which decreases the relative β-galactosidase activity by 5%, and the other of which decreases the β-galactoside permease activity by 5%. For the β-galactosidase mutation, $w = (0.004/0.95 + 0.130 + 0.866)^{-1} = 0.99979$, which corresponds to a selection coefficient of $s = 0.00021$; whereas for the β-galactoside permease mutation, $w = (0.004 + 0.130/0.95 + 0.866)^{-1} = 0.99320$, or $s = 0.00680$. This is about a 32-fold difference in the selection coefficient for an identical 5% decrease in activity. The double mutant has a selection coefficient of $s = 0.0070$, which is close to the sum of the individual selection coefficients.

Overall fitness is not a function of the activity of a single enzyme except under special circumstances, but more complex systems may also exhibit a pattern of increase in fitness that exhibits diminishing returns. One complex model that has considerable intuitive appeal likens the degree of genetic adaptation to the closeness with which a point in n-dimensional space approaches a fixed point representing "perfect" adaptation (Fisher 1930; Barton 1998). A mutation is regarded as a vector of random length shooting off in a random direction, which is beneficial if its point is closer to the fixed point and deleterious otherwise. The probability of fixation of a new mutation is governed by selection and RGD according to Equation 2.44 (Hartl and Taubes 1996, 1998; Orr 1998). The diminishing-returns aspect of the model emerges most clearly from the analysis by Orr (1998), who shows that, on the aver-

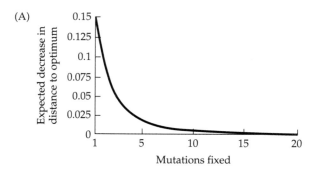

(A)

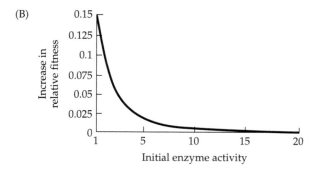

(B)

Figure 3.10 Expected change in relative fitness for successively fixed favorable mutations, for $n = 50$ dimensions. (A) Fisher's geometric model (Orr 1998). (B) Relative fitness as a function of β-galactosidase activity in *E. coli* (Hartl et al. 1985).

age, for each new mutation that is fixed the proportionate reduction in distance to the point representing perfect adaptation is given by

$$E(\text{proportionate reduction in distance to optimum}) \approx 1 - \frac{8}{3n}\sqrt{\frac{2}{\pi}} \qquad (3.24)$$

where n is the number of dimensions. This curve is plotted in Figure 3.10A for $n = 50$ dimensions, where the geometric decrease in the contribution of each successive fixed mutation is evident. A virtually identical curve is shown in 3.10B, which is the expected increase in relative fitness for each unit increase in enzyme activity, calculated from Equation 3.23.

MOLECULAR PHYLOGENETICS

Aligned amino acid or nucleotide sequences can also be used to make inferences about the ancestral relations between them. Making such inferences is the business of **molecular phylogenetics** (also called **molecular systematics**). Each aligned sequence yields a gene tree like that in Figure 3.1, but with sequences from different species and usually many more of them. A gene tree is not necessarily congruent with a species tree because of the way polymorphisms in ancestral species may become sorted in the descendant species. Figure 3.11 shows the species tree for seven species, S_1–S_7, and a nucleotide

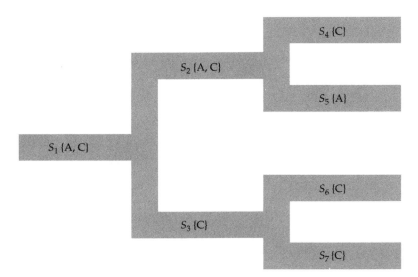

Figure 3.11   Inconsistency between a species tree and a gene tree resulting from random fixation of an ancestral polymorphism. Only one representative nucleotide site is shown.

site that is polymorphic for A and C in the common ancestor S_1. The polymorphism is retained in species S_2, but fixation occurs in all of the others. Because of the way the fixations occurred, this particular nucleotide site suggests that species S_4 is more closely related to S_6 and S_7 than it is to S_5, but in fact just the opposite is true. This kind of problem is the most acute for closely related species.

For more distantly related species there is a different kind of problem. It is that two or more independent mutations may occur at the same site (**multiple hits**). Because of multiple hits, two sites that differ may have undergone more than one change. There is also the possibility of **homoplasy,** which in the context of molecular phylogenetics refers to amino acid or nucleotide sites that are identical not because of identity by descent from a common ancestor but because of the following types of mutations:

- Parallel mutations at the same site (for example, two independent C → T substitutions)
- Convergent mutations at the same site (for example, C → T in one sequence, A → T in another)
- Reverse mutations at the same site (for example, C → T and then, later in time, T → C)

Because of multiple hits, the number of differences between two aligned sequences may underestimate the true number of changes that occurred. Some of the more straightforward methods of correcting for multiple hits are examined next.

The Multiple-Hit Problem

Amino acid replacements. Consider first the amino acid divergence of a protein in two related species. Suppose that the sequences were identical at the time of speciation, and that amino acid replacements become fixed independently in the two species at a rate λ per unit time. If λ is not too large, then at any time t after speciation, the expected proportion of amino acids that differ between the proteins, D_t, will be given by

$$D_t = (1 - D_{t-1})(2\lambda) + D_{t-1} \tag{3.25}$$

The rationale for this equation is that $(1 - D_{t-1})(2\lambda)$ is the proportion of sites, previously identical, in which one or the other protein underwent an amino acid replacement during the time interval in question, which must be added to the proportion of already existing differences D_{t-1} in order to give the total. The factor of 2 is present because the total time for evolution is $2t$ units (t units in each lineage after the species split).

Subtracting 1 from both sides of Equation 3.25 puts it in the form $D_t - 1 = (D_{t-1} - 1)(1 - 2\lambda)$, which can be solved by successive substitutions. Since $D_0 = 0$, the solution is

$$D_t = 1 - (1 - 2\lambda)^t \approx 1 - e^{-2\lambda t} \qquad (3.26)$$

Because λ is the rate (probability) of amino acid replacement per unit time, the expected number of replacements per amino acid site between two sequences at any time t is $K_a = 2\lambda t$, whereas the observed proportion of replacements is D_t. Substituting K_a into Equation 3.26 and rearranging leads to the estimate $\langle K_a \rangle$ and its variance

$$\langle K_a \rangle = -\ln(1 - D) \qquad \mathrm{Var}\langle K_a \rangle = \frac{D}{L(1 - D)} \qquad (3.27)$$

where $D = D_t$ to get rid of the subscript and L is the number of amino acids in the protein. If t is known, then λ can be estimated from the relation $\lambda = K_a/2t$. The estimate of K_a is used in preference to D in estimating the rate of molecular evolution because K_a includes the correction for multiple mutational hits. For relatively short intervals of evolutionary time, during which multiple substitutions remain uncommon, the correction is minor. For both amino acid replacements and nucleotide substitutions, a rule of thumb is that when $D \leq 10\%$ the correction for multiple hits is small, because when $D \leq 10\%$ the probability of two independent hits at the same site is $\leq 10\% \times 10\% = 1\%$.

Nucleotide substitutions. Multiple hits are a more serious problem for nucleotide sites than for amino acid sites. The reason is that, with only four possible nucleotides, homoplasy due to parallelism, convergence, or reversal is more likely. The correction for this possibility requires some inevitably simplifying assumptions about the mutational process. The simplest model is the **one-parameter model** (Jukes and Cantor 1969), which assumes that any nucleotide is equally likely to mutate and be substituted by any other nucleotide. For this model, the analog of Equation 3.27 is

$$\langle K_n \rangle = -\left(\frac{3}{4}\right)\ln\left(1 - \frac{4}{3}D\right) \qquad \mathrm{Var}\langle K_n \rangle = \frac{D(1 - D)}{L\left(1 - \frac{4}{3}D\right)^2} \qquad (3.28)$$

where K_n is the expected number of nucleotide substitutions per nucleotide site, D is the observed proportion of differences, and L the total number of nucleotides in the sequence. Note that D goes to $3/4$ as K_n becomes large, which means that two sequences are expected to differ at 75% of their nucleotide sites if they have evolved independently for so long that each site has been substituted multiple times.

Because multiple hits are potentially a serious problem for analyzing sequences that are very divergent, considerable effort has gone into corrections based on more realistic models than that in which all nucleotide substitutions are equally likely. There is the **two-parameter model** (Kimura 1980), which ascribes different mutation rates to transitions and transversions. A **transition** is a change from one pyrimidine nucleotide (T or C) into the other, or from one purine nucleotide (A or G) into the other; a **transversion** is a change from a pyrimidine to a purine or the other way around. Models with 4, 6, 9, and 12 parameters have also been investigated. (The 12-parameter model assigns a different substitution rate to each of the four nucleotides changing into any of the other three.) Mathematical details of the models can be found in Zharkikh (1994) and a good general discussion in Li (1997). In each case the substitution parameters are estimated from the data themselves. Even the 12-parameter model ignores differences in substitution rate due to effects of nearest neighboring nucleotides or local base composition of the genome, but such effects may be important (Bernardi 1995; Yang et al. 1998).

Phylogenetic Inference

Molecular systematics has revolutionized approaches to the biological classification of organisms because it uses data that are independent of morphology. The ancestral relations between organisms inferred from molecular sequences usually support those inferred from morphological characters (Patterson et al. 1993). Most of the discrepancies involve species with morphological differences that are few in number, inconsistent, or inconclusive. A case in point is the ancestral relationship between the gorilla, chimpanzee, and human. Unresolved on morphological grounds, the molecular data strongly favor the gorilla as the first to branch off from the common ancestor, but even so about 30 kb of aligned sequence was required to reach statistical significance (Patterson et al. 1993).

Many methods have been developed to infer the ancestral relations among a set of aligned sequences. They can be compared by the analysis of phylogenetic trees obtained either through computer simulations of sequence evolution (Nei 1996) or from real organisms when the true phylogeny is known, for example from experiments (Hillis et al. 1994). The methods differ in their

- efficiency in the use of computer time and the number of sequences that can be analyzed.

- power in identifying the correct tree for a given amount of data.
- consistency in identifying the correct tree with increasing probability as the amount of data increases.
- robustness in identifying the correct tree even when some of the assumptions of the method are in error.

Not surprisingly, all methods perform reasonably well if the data conform to the underlying assumptions of the method and if there are sufficient data. The most important factor in consistency seems to be the accuracy of the correction for multiple hits that is adopted (Penny et al. 1993). Even then, most methods give a disappointing performance when the rate of evolution varies dramatically from one branch to the next (Nei 1996).

Because no one method is superior by every criterion under all conditions, a variety of methods coexist. Many authors choose to analyze their data using multiple methods in hope that the resulting trees will differ at most in nonessential details. A detailed discussion of the methods and their relative merits and deficiencies is beyond the scope of this book but can be found in Hillis and Moritiz (1990), Hillis et al. (1994), Nei (1996) and Li (1997). The most commonly used methods can be classified under three broad headings.

- **Distance methods,** based on the pairwise differences between the sequences, corrected for multiple hits. These include:
  - *Unweighted pair-group method with arithmetic mean (UPGMA),* which is everybody's favorite whipping post because it assumes a constant rate of evolution in each branch and performs poorly when this assumption is violated (as it often is).
  - *Minimum evolution,* which examines every possible tree and selects the one that minimizes the total branch lengths. This approach is computationally intractable for a large number of sequences because there are so many possible trees, namely, $(2n - 5)!/[2^{n-3}(n - 3)!]$ unrooted bifurcating trees for n sequences.
  - *Neighbor joining,* which sequentially groups the most closely related pairs of sequences. This method is extremely efficient computationally and usually yields trees close to the minimum evolution tree.

- **Parsimony methods,** which systematically search among possible trees to identify that with the minimum number of mutational steps.
  - *Unweighted parsimony* treats each type of change (for example, transition or transversion) as equally informative.
  - *Weighted parsimony* gives some types of changes (usually transversions) greater importance in selecting the best tree. Weighted parsimony usually performs better than unweighted (Hillis et al. 1994).

- **Maximum likelihood,** which assumes a model of nucleotide or amino acid substitution and, based on this model, identifies the tree that maximizes the probability of obtaining the observed sequences. Intuitively appealing but computationally intensive, this method is quite tolerant of violation of its assumptions and performs quite well even when substitution rates differ moderately in different branches (Yang 1994, 1996).

Molecular Clocks

In spite of variation in rates of evolution among different molecules, the average rate of molecular evolution often manifests an approximate uniformity throughout long periods of evolutionary time. Such uniformity is known as a **molecular clock.** An example of clocklike uniformity in amino acid substitutions is illustrated in the gene tree of the α-globin gene depicted in Figure 3.12, which was obtained by neighbor joining from the distance data (K) below the

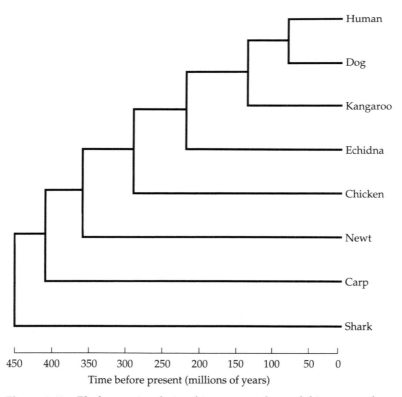

Figure 3.12 Phylogenetic relationships among the α-globin genes of selected vertebrate species and the approximate times of species divergence. (After Kimura 1983.)

Table 3.1 Rate of evolution in the α-globin gene

	Shark	Carp	Newt	Chicken	Echidna	Roo	Dog	Human
Shark		59.4	61.4	59.7	60.4	55.4	56.8	53.2
Carp	0.90		53.2	51.4	53.6	50.7	47.9	48.6
Newt	0.95	0.76		44.7	50.4	47.5	46.1	44.0
Chicken	0.91	0.72	0.59		34.0	29.1	31.2	24.8
Echidna	0.93	0.77	0.70	0.42		34.8	29.8	26.2
Roo	0.81	0.71	0.64	0.34	0.43		23.4	19.1
Dog	0.84	0.65	0.62	0.37	0.35	0.27		16.3
Human	0.76	0.67	0.58	0.28	0.30	0.21	0.18	
Average K	0.87	0.71	0.63	0.35	0.36	0.24	0.18	
Time	450	410	360	290	225	135	80	

Source: Data from Kimura 1983.

Note: Values above the diagonal are the observed percent amino acid differences D per pairwise comparison; those below are the values K corrected for multiple hits. The average values of K and estimated times of divergence (in millions of years) are given at the bottom. Roo means kangaroo. (Data from Kimura 1983.)

diagonal in Table 3.1. The numbers above the diagonal are the percent amino acid differences (D) between the α-globin sequences, and the K values were obtained using Equation 3.27. For example, the α-globins of dog and human differ in 23 of 141 amino acid sites, or D = 16.3%, which yields $\langle K \rangle$ = $-\ln(1 - 0.163) = 0.178$, rounded off in the table to 0.18. The percentages exclude differences that result from the insertion or deletion of amino acids, which are called **gaps** in sequence comparisons. For example, the comparison between human and shark alpha globin is based on 139 aligned amino acids and excludes gaps that span 11 amino acid sites. Although gaps are not typically used in phylogenetic analysis, they can contain a great deal of information about the number and size distribution of insertions and deletions and the evolution of genome size (Petrov et al. 1996, 1999).

Table 3.1 also includes the average value of $\langle K \rangle$ in all comparisons with humans and the estimated divergence times based on paleontological data. The average number of hits per site is plotted against divergence time in Figure 3.13. The very close fit to a straight line is evident. Since the divergence time is half of the total time available for evolution (because each divergence creates two independently evolving branches), the rate of evolution λ can be estimated as one-half times the slope of the line in Figure 3.13. For these data, the slope is 1.8×10^{-9}, and therefore $\langle \lambda \rangle = 0.9 \times 10^{-9}$ amino acid substitutions per amino acid site per year. The good fit of the points to the straight line indicates that the average rate of α-globin evolution has been

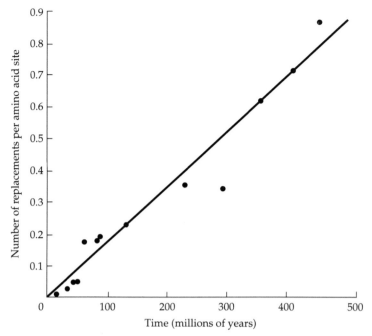

Figure 3.13 Relation between estimated number of amino acid substitutions per amino acid site in the α-globin gene between pairs of the vertebrate species, plotted against estimated time of species divergence. The straight line is expected based on a uniform rate of amino acid substitution during the entire period. (After Kimura 1983.)

approximately constant for the past 450 million years, but different genes, and different parts of genes, evolve at different rates (Figure 3.3). The α-globin gene is near the middle of the spectrum.

Strictly speaking, the molecular "clock" is not very clocklike. The rate of gene substitutions differs in different lineages as well as in different molecules. It is at least twofold faster in rodents than in higher primates, and about 30% faster in the Old World monkey lineage than in the human lineage (Li 1997). The rate of evolution even differs significantly in certain lineages of insects (Petrov et al. 1999). These differences are thought to be due to differences in mutation rate per year as a function of generation time. Furthermore, a molecular "clock" for the evolution of a molecule can have a variance in substitution rate very much larger than that expected from the Poisson distribution in Equation 3.16, whose mean λ and variance are equal. Gillespie (1989) has found that nonsynonymous substitutions in mammals have a variance that is about seven times greater than the mean, which he attributes to evolutionary episodes in which the rate of amino acid replacement is accelerated. This pattern has not been observed for nonsynonymous substitutions in *Drosophila*, but it does seem to hold for synonymous substitutions, perhaps due to selection for optimal codon usage (Zeng et al. 1998).

TRANSPOSABLE ELEMENTS

Most species of organisms contain DNA sequences called **transposable elements (TEs)** that can move from one location in the genome to another, some via a DNA intermediate and others via an RNA intermediate. Because of their ability to replicate and transpose, TEs can increase in copy number in the genome. Many TEs are also capable of **horizontal transmission** between reproductively isolated species (Robertson 1993; Robertson and MacLeod 1993; Clark et al. 1994; Capy et al. 1994; Hagemann et al. 1996). In some cases, horizontal transmission is evident from close sequence similarity of TEs in two otherwise very distantly related species (Garcia-Fernàndez et al. 1995; Robertson and Lampe 1995). For example, *Drosophila erecta* and the cat flea *Ctenocephalides felis* are very distantly related species, as evidenced by the 40% nucleotide identity found in the coding sequence for their sodium–potassium transmembrane pump, a generally very conserved gene. Yet both species contain a transposable element called *mariner* that shows 96–99% nucleotide identity (Lohe et al. 1995). The vectors of horizontal transmission of TEs have not been identified, nor is the rate of horizontal transmission well estimated.

Models for the population dynamics of transposable elements usually incorporate several features (Langley et al. 1983; Sawyer and Hartl 1986; Charlesworth et al. 1994):

- A rate of infection, in which genomes previously lacking the transposable element become infected with it.
- A rate of transposition, which determines how rapidly the copy number increases; the effects of regulation are taken into account by assuming that the rate of transposition is a decreasing function of copy number.
- A mechanism, or combination of mechanisms, for eliminating elements from the population; otherwise, the copy number would increase indefinitely. The usual assumption is that the presence of transposable elements in the genome decreases fitness. This assumption is consistent with the inference that most TE insertions are deleterious, since each individual insertion in a population remains rare and is eliminated rather quickly (Golding et al. 1986). There is also direct evidence that transposition of the *mariner* element in somatic cells decreases lifespan (Nikitin and Woodruff 1995). To suggest that most TE insertions are deleterious is not to deny that some TE insertions have acquired important functions in

evolution (Britten 1997; Miller et al. 1997). But TEs seem to persist mainly because they are efficient molecular parasites of the genome, not because a few individual insertions have opportunistically been put to use.

Insertion Sequences and Transposons in Bacteria

Bacteria contain several types of transposable elements, the simplest of which are **insertion sequences,** typically 1–2 kb in length containing short nucleotide sequences repeated at each end in inverted orientation and at least one long, open reading frame coding for the transposase protein that catalyzes transposition. The intestinal bacterium *E. coli* contains six well-characterized insertion sequences, the numbers of which have been determined among 71 natural isolates (Sawyer et al. 1987). These distributions have been used to estimate the parameters in population models.

Models of transposable elements in bacteria are greatly simplified because of asexual reproduction, a low rate of recombination among strains, and a low rate of deletion of insertion sequences. The "state" of a strain may be described in terms of the number of copies i of an element it contains. The parameters that govern the population genetics of a bacteria TE are:

- u, the rate at which uninfected cells become infected (that is, go from state $i = 0$ to state $i = 1$).
- T, the rate of transposition in infected strains (that is, the probability that a cell in state i goes to state $i + 1$ in the interval of one generation).
- s, the selection coefficient measuring the decrease in fitness of infected cells, relative to uninfected cells.

The most general models of this type allow T and s to be functions of i, but here we will assume that they are constant. However, assuming a constant value for T does incorporate regulation of transposition, because the rate of transposition *per element* in a strain in state i equals T/i, which is a decreasing function of i.

With these assumptions, the population attains an equilibrium distribution of numbers of elements, in which the probability p_i that a strain contains exactly i elements is given by

$$p_0 = \alpha \quad \text{and} \quad p_i = (1-\alpha)(1-\Theta)\Theta^{i-1} \quad (i \geq 1) \tag{3.29}$$

where $\alpha = 1 - (u/s)$ and $\Theta = T/(T + s - u)$ (Sawyer and Hartl 1986; Sawyer et al. 1987).

Equation 3.29 may be applied to the concrete case of insertion sequence IS30 in *E. coli*, in which the distribution of numbers among 71 strains fits a model with $\alpha = 1/2$ and $\Theta = 1/2$ (Sawyer et al. 1987). With these parameters the expected distribution reduces to the remarkably simple formula $p_i = (1/2)^{i+1}$ for $i \geq 0$. Among 71 natural isolates, the observed and expected numbers containing 0, 1, 2, 3, 4, and ≥ 5 elements were

Observed number	36	16	13	2	2	2
Expected number	35.5	17.8	8.9	4.4	2.2	2.2

This is obviously a very good fit, even though α and Θ were estimated from the data.

Aside from their own evolutionary dynamics, insertion sequences are important because they can mobilize other sequences in the genome. When two copies of an insertion sequence are on flanking sides of an unrelated sequence, the inverted repeats that are used in transposition are preferentially those at the extreme ends, creating a composite transposable element or **transposon** that transposes as a single unit. For example, the transposon *Tn5* consists of a central DNA sequence containing three genes conferring resistance to the antibiotics neomycin, streptomycin, and bleomycin, which are flanked by copies of the insertion sequence *IS50* enabling mobilization of *Tn5* as an intact unit (Blot et al. 1994). Antibiotic-resistance transposons are also important building blocks of infectious bacterial plasmids called **resistance transfer factors** that contain multiple antibiotic-resistance genes (Kruse and Sorum 1994).

Transposable Elements in Eukaryotes

Long-term persistence of TEs is possible in asexual organisms owing to transmissible plasmids and other mechanisms that allow them to be transmitted from one cell to the next (Hartl and Sawyer 1988). Such mechanisms are not necessary in sexual organisms. Sexual reproduction itself enables TEs to spread because the sexual process brings different genomes into physical association (Charlesworth 1985; Charlesworth and Langley 1989; Charlesworth et al. 1994). As in asexual populations, an indefinite increase in TE copy number is checked in part by regulatory mechanisms that have evolved at least in some TEs that reduce the net rate of transposition as copy number increases (Engels 1997; Hartl et al. 1997). Equally as important are selective effects, which need not be very large to offset even unregulated transposition. If TEs are distributed at random among the genomes in a population, and if the relative fitness of an organism decreases with the number i of copies in its genome, then at equilibrium the average fitness $\bar{w}$ of the population, relative to a population lacking TEs, is given by

$$\bar{w} = e^{-\bar{i}T} \tag{3.30}$$

where $\bar{i}$ is the mean number of copies per individual and T is the transposition rate (Charlesworth 1985). Even in *Drosophila* the average number of TEs per genome is not known, but is thought to be on the order of 500, comprising about 50 different families of TEs averaging about 10 copies per family (Finnegan and Fawcett 1986). The transposition rate is also uncertain but may usually be on the order of 10^{-4} to 10^{-5}. These values would put $\bar{w}$ in the range 0.95–0.995, yielding a very rough estimate of the fitness cost per copy of 10^{-4} to 10^{-5}, prompting Charlesworth (1985) to remark that there is no great difficulty in explaining the maintenance of multiple families of TEs in a species by natural selection offsetting transposition.

FURTHER READINGS

Avise, J. C. 1994. *Molecular Markers, Natural History and Evolution.* Chapman and Hall, New York.

Chakravarti, A. 1984. *Human Population Genetics.* Van Nostrand Reinhold, New York.

Crow, J. F. and M. Kimura. 1970. *An Introduction to Population Genetics Theory.* Harper & Row, New York.

Ewens, W. J. 1979. *Mathematical Population Genetics.* Springer-Verlag, Berlin.

Gale, J. S. 1990. *Theoretical Population Genetics.* Unwin Hyman, London.

Golding, B. (ed.). 1994. *Non-Neutral Evolution: Theories and Molecular Data.* Chapman and Hall, New York.

Graur, D. and W.-H. Li. 2000. *Fundamentals of Molecular Evolution*, 2nd Ed. Sinauer Associates, Sunderland, MA.

Hartl, D. L. and A. C. Clark. 1997. *Principles of Population Genetics*, 3rd Ed. Sinauer Associates, Sunderland, MA.

Hillis, D. M., C. Moritz and B. K. Mable (eds.). 1996. *Molecular Systematics,* 2nd Ed. Sinauer Associates, Sunderland, MA.

Jacquard, A. 1978. *Genetics of Human Populations.* Trans. by D. M. Yermanos. Jones & Bartlett, Boston.

Kimura, M. 1983. *The Neutral Theory of Molecular Evolution.* Cambridge University Press, Cambridge.

Kimura, M. and T. Ohta. 1971. *Theoretical Aspects of Population Genetics.* Princeton University Press, Princeton, NJ.

Kingman, J. F. C. 1980. *Mathematics of Genetic Diversity.* Society for Industrial Applied Mathematics, Philadelphia.

Mitton, J. F. 1997. *Selection in Natural Populations.* Oxford University Press, New York.

Nei, M. 1987. *Molecular Evolutionary Genetics.* Columbia University Press, New York.

Nei, M. and S. Kumar. 2000. *Molecular Evolution and Phylogenetics.* Oxford University Press, New York.

Ohta, T. and K. Aoki. 1985. *Population Genetics and Molecular Evolution.* Springer-Verlag, New York.

Takahata, N. and A. G. Clark (eds.). 1993. *Mechanisms of Molecular Evolution.* Sinauer Associates, Sunderland, MA.

Weir, B. S. 1996. *Genetic Data Analysis II.* Sinauer Associates, Sunderland, MA.

PROBLEMS

The data in the accompanying table are configurations of synonymous polymorphisms in five alleles of the *bride of sevenless* (*boss*) gene of *D. melanogaster*. In addition to the polymorphic sites there were 433 monomorphic synonymous sites. The data are from Ayala and Hartl (1993).

Allele	Polymorphic site														
	1	2	3	4	5	6	7	8	9	10	11	12	13	14	15
a	T	C	T	C	C	T	G	T	A	G	T	G	C	C	G
b	T	C	C	T	C	C	A	C	G	A	G	G	C	T	G
c	C	A	T	T	C	C	A	C	A	A	T	T	T	C	G
d	C	A	T	C	T	C	A	C	A	G	T	T	C	T	A
e	C	A	T	C	T	C	A	C	A	G	T	T	T	C	A

3.1 For the *boss* data in the table:

  a. Calculate the number of pairwise mismatches at each polymorphic site.
  b. Classify the sample configuration at each site.
  c. Classify each site as to whether it is or is not phylogenetically informative.

3.2 For the *boss* data in the table, estimate the value of θ from:

  a. The nucleotide polymorphism S, and give the standard deviation of the estimate assuming no recombination.
  b. The nucleotide diversity π, and give the standard deviation of the estimate assuming no recombination.

3.3 For the *boss* data in the table, calculate Tajima's D statistic and interpret the result.

3.4 For a single polymorphic nucleotide site in a sample of 12 aligned sequences, how many pairwise comparisons are possible? For each of the situations below, specify: (1) the minimum number of pairwise mismatches and the sample configuration that yields it, and (2) the maximum number of pairwise mismatches and the sample configuration that yields it.

  a. The site is occupied by two nucleotides.
  b. The site is occupied by three nucleotides.
  c. The site is occupied by four nucleotides.

3.5   The first 18 amino acids present at the amino terminal end of the human and mouse immune γ-interferon proteins constitute a signal peptide that is used in secretion of the molecules. Calculate the proportion of amino acids that differ in the two signal sequences.

> Human: M-K-Y-T-S-Y-I-L-A-F-Q-L-C-I-V-L-G-S
> Mouse: M-N-A-T-H-C-I-L-A-L-Q-L-F-L-M-A-V-S

Use these data to estimate the average rate of amino acid replacement in the signal peptide of γ interferon during the divergence of mice and humans. Based on fossil evidence, the separation of these species took place approximately 80 million years ago.

3.6   The evolutionary separation of kangaroos and dogs took place approximately 135 million years ago, and in the evolution of vertebrates as a whole the α-hemoglobin molecule has been undergoing amino acid replacement at the average rate of $\lambda = 1 \times 10^{-9}$ amino acid replacements per amino acid site per year. Calculate the proportion of amino acid sites that are expected to differ in the α-hemoglobin of kangaroo and dog.

3.7   The following 60 nucleotides are found in the coding region of the *trpA* genes in strains of the related enteric bacteria *Escherichia coli* strain K12 and *Salmonella typhimurium* strain LT-2. The *trpA* gene encodes one of the subunits of the enzyme tryptophan synthetase used in the synthesis of tryptophan.

```
K12:  V  A  P  I  F  I  C  P  P  N  A  D  D  D  L  L  R  Q  I  A
K12:  GTCGCACCTATCTTCATCTGCCCGCCAAATGCCGATGACGACCTGCTGCGCCAGATAGCC
LT2:  ATCGCGCCGATCTTCATCTGCCCGCCAAATGCGGATGACGATCTTCTGCGCCAGGTCGCA
LT2:  I  A  P  I  F  I  C  P  P  N  A  D  D  D  L  L  R  Q  V  A
```

a. Estimate the amount of nucleotide divergence K_n and amino acid divergence K_a and their standard deviations.

b. Assuming that *Escherichia* and *Salmonella* diverged at around the time of the mammalian radiation 80 million years ago, estimate the rates of nucleotide substitution and amino acid replacement.

3.8   In the region of the *E. coli trpA* gene given above:

a. Use the genetic code in Table 1.1 to assign a degeneracy of 0, 2, or 4 to each nucleotide, considering isoleucine codons as twofold degenerate.

b. For each difference between *E. coli* and *S. typhimurium,* consider the difference as synonymous if either the site is fourfold degenerate or else if it is twofold degenerate and the change is a transition (that is, A to G or the reverse; or T to C or the reverse). Consider the difference as nonsynonymous if either the site is nondegenerate or else if it is twofold degenerate and the change is a transversion (that is, A or G to T or C).

c. Use Equation 3.28 to calculate the proportion of nonsynonymous nucleotide substitutions per nonsynonymous site and the proportion of synonymous substitutions per synonymous site.

3.9 The mitochondrial DNA molecules (mtDNA) of 21 humans of diverse geographic and racial origin were digested with 18 restriction enzymes, 11 of which exhibited one or more fragments in which size polymorphism occurred (Brown 1980). All fragment polymorphisms could be explained by single-nucleotide differences; thus, there was no evidence for insertions, deletions, or other mtDNA rearrangements. Altogether, 868 nucleotide sites were assayed for differences among individuals, and the average number of differences per nucleotide site per individual was estimated at 0.0018. Assuming that mammalian DNA undergoes sequence divergence at the rate of $5–10 \times 10^{-9}$ nucleotide substitutions per site per year, and that the rate is uniform in time, estimate the length of time since all of the 21 contemporary mtDNA molecules last shared a common ancestor.

3.10 In populations of *E. coli* undergoing competition for lactose, the fitness of a mutant strain, relative to the standard *E. coli* K12 strain, is given approximately by $1/w = (0.004/Z) + (0.130/Y) + 0.866$, where Z is the activity of the β-galactosidase and Y is that of the β-galactoside permease, both measured relative to their counterparts in the wildtype K12 strain (Dykhuizen et al. 1987).

    a. What is the decrease in relative fitness of a strain with a mutant permease allele that decreases permease activity by 20%?

    b. What magnitude of change in β-galactosidase activity would be required to cause the same decrease in fitness?

    c. What is the predicted relative fitness of the double mutant?

3.11 In a region of 4146 nucleotides in the $^{A}\gamma$ and $^{G}\gamma$ fetal globin genes, which arose by gene duplication followed by sequence divergence, there were 34 "gaps" in the aligned sequences, each resulting from an insertion or deletion event (indel). The length distribution of the gaps, in number of nucleotides, was as follows:

Length	No.	Length	No.	Length	No.	Length	No.	Length	No.
1	12	2	3	3	2	4	4	5	6
6	1	7	2	8	0	9	0	10	1
13	1	23	1	122	1				

    a. Assuming that each indel results from a single insertion/deletion event, use Equation 3.27 to estimate K for indels in the sequences and the standard deviation.

    b. Assuming a divergence time of 34 million years between the sequences, what is the rate of incorporation of indels?

    c. How does this number compare with the rate of 1% nucleotide sequence divergence per 2.2 million years?

3.12 A 300-bp highly repetitive sequence in the human genome is present in more than 300,000 copies and accounts for about 3% of the total DNA. The sequence is known as *Alu* because of an *Alu*I restriction site it contains. Two randomly chosen *Alu* sequences differ, on the average, at 15–20% of their nucleotide sites.

Assuming a rate of sequence evolution of 5 nucleotide substitutions per nucleotide site per 10^9 years (approximately the rate for pseudogenes), estimate the average time of divergence of two randomly chosen *Alu* sequences.

3.13 The distribution of copy number of several types of transposable insertion sequences has been studied in 71 natural isolates of *E. coli,* yielding the data shown below (Sawyer *et al.* 1987).

Type of element		Number of copies					
		0	1	2	3	4	≥ 5
IS1	No. strains	11	14	8	6	7	25
IS2	No. strains	28	8	12	5	5	13
IS4	No. strains	43	5	5	3	5	10

Equation 3.29 can be fit to these data to estimate α and Θ. Then a chi-square test can be used to assess goodness of fit to the model. Calculate the expected numbers for each of the models below and carry out a chi-square test. (Each test has three degrees of freedom because two parameters were estimated from the data.)

    a. *IS1* with $\alpha = 1/5$ and $\Theta = 5/6$.
    b. *IS2* with $\alpha = 2/5$ and $\Theta = 2/3$.
    c. *IS4* with $\alpha = 2/3$ and $\Theta = 3/4$.

3.14 Polymorphism and divergence were studied among synonymous and replacement sites in the *bride of sevenless* (*boss*) gene in samples from four species of *Drosophila* (Ayala and Hartl 1993).

    a. In these samples there were 71 fixed and 106 polymorphic synonymous differences as well as 8 fixed and 13 polymorphic replacement differences. Carry out a chi-square test of polymorphism and divergence and interpret the result.

    b. Among the species that had also been examined for the alcohol dehydrogenase gene (*Adh*) by McDonald and Kreitman (1991), there were 60 fixed and 75 polymorphic synonymous differences at *boss* as compared with 20 fixed and 44 polymorphic synonymous differences at *Adh*. Carry out a chi-square test to determine whether there is a significant difference in polymorphism and divergence at synonymous sites between *boss* and *Adh*.

3.15 Equation 3.15 gives the expected number of polymorphic sites with exactly r mutant and $n - r$ nonmutant nucleotides for selectively neutral substitutions in a Poisson random field. Note that ratios of sums or differences of λ_r are independent of $2N\mu k_s$ because this factor cancels in the numerator and denominator. Defining $I(r)$ as the integral in Equation 3.15 [that is, $I(r) = \lambda_r/(2N\mu k_s)$], numerical integration for $n = 5$ yields $I(1) = 1.00$, $I(2) = 0.50$, $I(3) = 0.33$, and $I(4) = 0.25$.

    a. Use these results to test whether the sample configurations of synonymous polymorphisms in the *bride of sevenless* (*boss*) gene of *D. melanogaster* fit a PRF model with $s = 0$. Among 27 synonymous polymorphisms in 5 alleles sequenced, 17 had a (4, 1) sample configuration and 10 had a (3, 2) sample configuration.

b. What ratio of (4, 1) : (3, 2) sample configurations would be needed to make the X^2 significant at the 5% level, given 27 polymorphic sites? How would a significant result be interpreted?

3.16 The infinite sites model of DNA sequence evolution assumes that each new mutation occurs at a nucleotide site not previously mutated. Watterson (1975) has shown that, in the absence of recombination, the probability that two sequences chosen at random will differ at exactly $K_2 = i$ sites is given approximately by $\dfrac{1}{1+\Theta}\left(\dfrac{\Theta}{1+\Theta}\right)^i$ for $i = 0, 1, 2, \ldots$, where $\Theta = 4NU$ and U is the aggregate mutation rate summed across all nucleotides in the sequence.

a. Calculate $\Pr\{K_2 = i\}$ for $\Theta = 1$ and $i = 0$–10.

b. In terms of Θ, what is the probability that two randomly chosen sequences are identical?

c. In terms of Θ, what is the probability that two randomly chosen sequences differ at one or more sites? Interpret this result in light of Equation 2.12.

d. Would you expect recombination to increase or decrease the probability of one or more mismatches? Explain why.

3.17 Equation 2.45 gives the average time to fixation of a new neutral mutation that is destined to be fixed as $4N$. This result also follows from coalescence theory, because it is the average time for the coalescence of all the allele lineages in a population. To see this for yourself:

a. Calculate $\displaystyle\sum_{i=2}^{N} \overline{T_i}$ where $\overline{T_i}$ is given by Equation 3.3.

b. Explain in words why this is the expected total time to coalescence of all allele lineages in the population.

c. Show that the sum calculated in part (a) is approximately $4N$ for N reasonably large.

Solutions to the problems, worked out in full, can be found at the website
www.sinauer.com/hartl/html

CHAPTER 4

The Genetic Architecture of Complex Traits

No study of population genetics is sufficient without consideration of traits influenced by multiple genetic and environmental factors, often called **complex traits.** The **genetic architecture** of a complex trait includes all of the genetic and environmental factors that affect the trait, along with the magnitudes of their individual effects and the magnitudes of any interactions among the factors. The genetic architecture of a complex trait depends not only on the trait but also on the particular population. The genetic architecture is affected by genotype frequencies, the distributions of environmental factors, and such biological properties as age and gender. The approaches described in this chapter enable the genes affecting a complex trait to be identified and positioned along a genetic map. The environmental factors are less easily sorted out.

TYPES OF COMPLEX TRAITS

In human populations, the most common medical disorders with a genetic component are complex traits, including coronary artery disease, asthma, peptic ulcer, schizophrenia, and many congenital abnormalities. In domesticated animals and plants, most traits of commercial interest are complex traits, such as egg production, milk production, and yield of grain. Even as modern meth-

151

ods of genetic engineering are applied to animal and plant improvement, complex traits continue to be important because most desirable traits result from interactions of multiple genes and environmental factors.

Complex traits are often called **quantitative traits** to distinguish them from traits that appear in discrete categories, like round versus wrinkled pea seeds. Three types of quantitative traits may be distinguished:

- **Metric traits,** which are measured on a continuous, uninterrupted scale, such as height or weight.
- **Meristic traits,** which are measured by counting and include such traits as litter size or number of bristles. When the number of possible phenotypes is large, there is little distinction between a meristic trait and a metric trait.
- **Threshold traits,** which are discrete in that they are either present or absent in any one individual, but which are complex in that the underlying risk or liability toward the trait is determined by multiple genetic and environmental factors.

Because complex traits are affected by multiple genetic factors, they are also often called **polygenic traits.** Typically, each of the multiple genes underlying a complex trait has the feature that the mean difference in phenotype between the alternative genotypes is relatively small in comparison with the total variance in phenotype in the population. These gene are sometimes called **polygenes,** but more commonly they are referred to as **quantitative trait loci** or **QTLs.**

PHENOTYPIC VARIATION

Complex traits differ in phenotype from one individual to the next. For quantitative and meristic traits, the phenotype of a particular individual is called its **phenotypic value.** An example of phenotypic variation for a meristic trait is illustrated in Figure 4.1. The trait is the number of bristles on an abdominal segment in *Drosophila*. For bristle number the phenotypic values typically range from 13–25 with a modal number of about 19. The smooth curve is a normal distribution approximating the histogram. Two parameters are of great interest for any quantitative trait:

- The **mean,** usually denoted μ, defined as the expected phenotypic value across all individuals in the population, $E(x)$.
- The **variance,** usually denoted σ^2, defined as the expected value of the squared deviation of each phenotypic value from the population mean, $E(x - \mu)^2$.

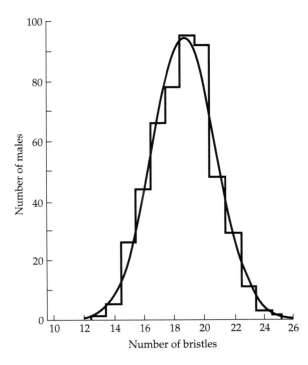

Figure 4.1 Number of bristles on the fifth abdominal sternite in 530 males of a strain of *D. melanogaster*. (Data courtesy of Trudy Mackay.)

These parameters can be estimated from the values observed in a random sample of individuals as follows:

$$\bar{x} = \langle\mu\rangle = \frac{\sum\limits_{i=1}^{n} x_i}{n} \qquad s^2 = \langle\sigma^2\rangle = \frac{\sum\limits_{i=1}^{n} (x_i - \bar{x})^2}{n-1} \tag{4.1}$$

In these formulas, x_i is the phenotypic value of the ith individual in a sample of n individuals. The symbols μ and σ^2 refer to the mean and variance in the whole population from which the sample was obtained, and $\langle\mu\rangle$ and $\langle\sigma^2\rangle$ denote estimates of μ and σ^2 based on the sample. The symbols $\bar{x}$ and s^2 are included because they are frequently used as symbols for the mean and variance of a sample. The use of $n-1$ in the denominator of s^2 is a correction for sampling error, and when n is reasonably large, the correction is minor. Another term frequently encountered is the **standard deviation,** which equals the square root of the variance. The standard deviation of the population as a whole is $\sqrt{\sigma^2} = \sigma$, whereas the standard deviation of the sample is

$\sqrt{s^2} = s$. For the abdominal bristle data in Figure 4.1, $\bar{x} = 18.74$ bristles and $s = \sqrt{4.306} = 2.075$ bristles.

The functional form of the normal distribution, denoted $N(\mu, \sigma)$, is defined by

$$f(x) = \frac{1}{\sqrt{2\pi\sigma}} e^{-\frac{(x-\mu)^2}{2\sigma^2}} \qquad (-\infty < x < \infty) \qquad (4.2)$$

where, in this case, $\mu = 18.74$ and $\sigma^2 = 4.306$. As usual, $\pi = 3.14159$ and $e = 2.71828$ are constants. Note that the theoretical range of x is minus infinity to plus infinity, but for these values of the mean and variance the expected proportion of negative phenotypic values is negligible (less than 1.8×10^{-19}).

The proportion of individuals in the population that have phenotypic values in the range x_1 to x_2 is given by the integral of Equation 4.2 between x_1 and x_2. In particular, the proportion of the population with phenotypic values within d standard deviations of the mean (that is, between $x_1 = \mu - d\sigma$ and $x_2 = \mu + d\sigma$) equals the integral of Equation 4.2 between $x_1 = \mu - d\sigma$ and $x_2 = \mu + d\sigma$. Conversely, the proportion of the population that has a phenotypic value either less than $x_1 = \mu - d\sigma$ or greater than $x_2 = \mu + d\sigma$ equals 1.0 minus this integral. Some representative values are shown in Figure 4.2. Since the normal distribution is symmetrical, half the deviations represent phenotypic values smaller than $x_1 = \mu - d\sigma$ and half represent phenotypic values larger than $x_2 = \mu + d\sigma$. For example, the value 0.317 in Figure 4.2 means that $0.317/2 \approx 16\%$ of a normal population has a phenotypic value less than one standard deviation from the mean; likewise, approximately 16% has a phenotypic value greater than one standard deviation from the mean. The remaining approximately 68% have phenotypic values that deviate by less then one standard deviation from the mean. The values for 1, 2, and 3 standard deviations in Figure 4.2 are the basis of the 68%, 95%, and 99.7% confidence intervals discussed in connection with allele-frequency estimates in Chapter 1 (Equations 1.1 and 1.2).

As a practical application we can use data for pupa weight in the flour beetle *Tribolium castaneum*, which is normally distributed with $\langle \mu \rangle = 2246.9$ mg and $\langle \sigma \rangle = 176.86$ mg (Enfield 1980). From the values for 1, 2, and 3 standard deviations in Figure 4.2 we can assert that $\approx 68\%$ of the pupae are in the weight range $\mu \pm \sigma$ (2070.0–2423.8 mg), $\approx 95\%$ are in the range $\mu \pm 2\sigma$ (1893.2–2600.6 mg), and $\approx 99.7\%$ are in the range $\mu \pm 3\sigma$ (1716.3–2777.5 mg).

The normal distribution occupies such a prominent place in the study of complex traits because of a statistical theorem called the central limit theorem, which states, roughly speaking, that the summation of many random, independent quantities conforms to the normal distribution. Since complex phenotypes are determined by multiple genetic and environmental factors,

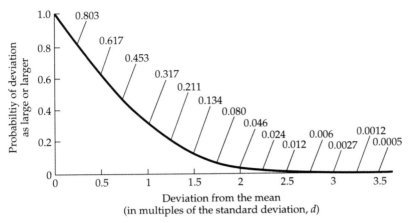

Figure 4.2 Probability that an observation taken from a normal distribution will deviate from the mean by a number of standard deviations as large or larger than some specified value d. The probabilities are given exactly by $\int_{-\infty}^{-d} N(0,1)dx + \int_{+d}^{+\infty} N(0,1)dx$, but there are quite accurate approximations. For deviations d between 0 and 2, use $1 - \sqrt{1 - e^{-2d^2/\pi}}$ and for $d > 2$ use $\left(\dfrac{0.7979}{d}\right)e^{-d^2/2}$.

a normal distribution is to be expected if the factors are independent and their effects approximately additive. Francis Galton (1822–1911), a cousin of Charles Darwin, spent a good deal of his life investigating applications of the normal distribution to phenotypic variation. He was sufficiently impressed by the wide applicability of the normal distribution to write (Galton 1889, p. 66):

> I know of scarcely anything so apt to impress the imagination as the wonderful form of cosmic order expressed by the "law of frequency of error" [the normal distribution]. Whenever a large sample of chaotic elements is taken in hand and marshaled in the order of their magnitude, this unexpected and most beautiful form of regularity proves to have been latent all along. The law would have been personified by the Greeks if they had known of it. It reigns with serenity and complete self-effacement amidst the wildest confusion. The larger the mob and the greater the apparent anarchy, the more perfect is its sway. It is the supreme law of unreason.

Although many quantitative and meristic traits are distributed normally, some are not. For example, data in percentages (p) are often not normally distributed, but they may become nearly so when expressed as $x = \arcsin\sqrt{p}$.

Another "normalizing transformation" useful in other cases is the natural logarithm.

GENETICS AND ENVIRONMENT

A major assumption of quantitative genetics is that genetic and environmental effects on complex traits are additive. This assumption cannot be verified explicitly because the individual factors are not directly observable. But it allows the phenotypic value of any individual to be written as the sum of three terms:

- The mean μ of the entire population
- A deviation from the population mean due to the genotype of the individual
- A deviation from the population mean due to the environment of the individual

The additivity assumption is written in symbolic form as

$$P_i = \mu + G_i + E_i \tag{4.3}$$

where P_i is the phenotypic value of the ith individual in the population, μ is the mean of the population, G_i is the deviation of the ith individual from the population mean due to genetic factors, and E_i is the deviation of the ith individual from the population mean due to environmental factors.

Genotypic Variance and Environmental Variance

Equation 4.3 implies that the variance in phenotypic value (the **phenotypic variance, σ_p^2**) can be partitioned into one component due to variation among the genetic factors, which is called the **genotypic variance, σ_g^2**, and another component due to variation among the environmental factors, which is called the **environmental variance, σ_e^2**. The basis of this partitioning is the definition of the variance in terms of the expected value of the squared deviations:

$$\sigma_p^2 = E(P_i - \mu)^2 = E(G_i^2) + E(E_i^2) = \sigma_g^2 + \sigma_e^2 \tag{4.4}$$

Strictly speaking there should also be a covariance term on the right-hand side of Equation 4.4, but its value is 0 if the genetic and environmental factors are independent.

The biological meaning of Equation 4.4 is shown for the alleles of one gene in Figure 4.3. The solid curves represent the phenotypic distributions in the three genotypes AA, AA', and $A'A'$, with their means denoted G_1, G_2, and G_3. The dashed curve represents the phenotypic distribution in the entire

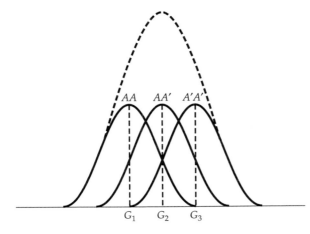

Figure 4.3 Overall phenotypic distribution (dashed curve) of a quantitative trait in a hypothetical randomly mating population, showing the underlying phenotypic distributions for three genotypes.

population. The phenotypic variance σ_p^2 is the variance of the dashed distribution. The genotypic variance σ_g^2 is the variance among the G's , which is $\sigma_g^2 = p^2 G_1^2 + 2pq G_2^2 + q^2 G_3^2$ where p is the allele frequency of A. Although the G's are not generally known, σ_g^2 must equal zero in a genetically uniform population. Hence the observed variance of a genetically uniform population provides an estimate of σ_e^2, whereas the observed variance of a randomly bred population provides an estimate of $\sigma_g^2 + \sigma_e^2$. An estimate of σ_g^2 can be obtained by subtraction, since $\sigma_g^2 = (\sigma_g^2 + \sigma_e^2) - \sigma_e^2$. An example with thorax length in *Drosophila* is shown in Table 4.1. In this case, genetic variation among flies in the randomly bred population accounts for about $0.180/0.366 = 49.2$ percent of the phenotypic variance in the population. This method of separating the genotypic and environmental variance has also been used in studies of monozygotic twins (identical twins) in human populations,

Table 4.1 Estimation of the genotypic variance (σ_g^2) and environmental variance (σ_e^2) of thorax length in *Drosophila melanogaster*[a]

	Populations	
Variance	**Random-bred**	**Uniform**
Theoretical	$\sigma_g^2 + \sigma_e^2$	σ_e^2
Observed	0.366	0.186

$\sigma_e^2 = 0.186$

$\sigma_g^2 = (\sigma_g^2 + \sigma_e^2) - \sigma_e^2 = 0.366 - 0.186 = 0.180$

Source: Data from Robertson (1957).
[a]In units of 10^{-2} mm^2.

because MZ twins have identical genotypes but different environmental experiences.

Broad-Sense Heritability

For a complex trait, one measure of the aggregate effect of all genetic factors combined is the **broad-sense heritability** H^2, defined as the ratio of the genotypic variance to the phenotypic variance:

$$H^2 = \frac{\sigma_g^2}{\sigma_p^2} = \frac{\sigma_g^2}{\sigma_g^2 + \sigma_e^2}$$

(4.5)

Hence, if $H^2 = 0$, all of the phenotypic variance is attributable to differences in environment, and if $H^2 = 1$, all of the phenotypic variance is attributable to differences in genotype.

Equation 4.4 is important in suggesting how genetic and environmental effects on the variance may be separated. What about effects on the mean phenotypic value? For example, suppose two human populations differ in average height and that the values of σ_g^2 and σ_e^2 are known in both populations. What can be inferred about the genetic versus environmental cause of the difference in the mean? Nothing whatever. At one extreme, the two populations could be quite similar genetically but their environments somewhat different. At the other extreme, the two populations could differ genetically and the environments be quite comparable. Without additional data, nothing more can be said. The variance components have a very restricted utility and apply only to differences in phenotype within populations, not between populations.

Genotype-Environment and Other Interactions

Equation 4.4 ignores several possible sources of nonindependence between genotype and environment.

Genotype-environment interaction (GEI). GEI occurs when the genotypic and environmental effects are not additive, but differ according to which genotype is in which specific environment. An example of genotype-environment interaction in maize is illustrated in Figure 4.4. The two strains are hybrids formed by crossing different pairs of inbred lines, and the index of environmental quality is based on soil fertility, moisture, and other factors. The overall mean yields of A and B, averaged across all environments, are both about 8500 kg/ha. However, A clearly outperforms B in the very stressful environments (negative), whereas B outperforms A in the very favorable environments (positive). A curve showing the phenotype of a genotype across the range of environments is called the **norm of reaction** for the geno-

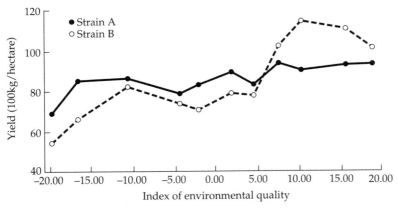

Figure 4.4 Genotype-environment interaction in two strains of maize. Each curve is the norm of reaction of the strain. (Data from Russell 1974.)

type. GEI is indicated when the norms of reaction cross (Fry et al. 1996). The important implication of GEI is that the deviations due to genotype are not independent of the deviations due to environment

Genotype-environment association (GEA). This is another cause of nonindependence of genotypic and environmental deviations. It occurs when the genotypes in a population are not distributed randomly in all the possible environments. With GEA it is difficult, if not impossible, to separate genetic and environmental causes of variation, because there is a systematic association of certain genotypes with certain environments. One example of a deliberate GEA is the practice of many dairy farmers to provide more feed supplements to cows that produce more milk, hence cows that have superior milk-yield genotypes are also provided a superior nutritional environment.

Genotype-by-sex interaction (GSI). GSI occurs when the magnitude of a genotypic deviation depends on the sex of the individual. It is a potential problem in genetic analysis because the expression of many complex traits depends in part on developmental, hormonal, or other factors associated with sex (Nuzhdin et al. 1997). An example is seen in human height, in which the female average is about 10 cm smaller than the male average. Yet there is no reason for thinking that height-related genotypes are distributed differently in females and males. In this case, considering a mixed population of females and males, there is a genotype-by-sex interaction because the majority of genotypes will yield a negative deviation from the overall average if the individual is a female, and a positive deviation if the individual is a male.

Genetic Effects on Complex Traits

To connect Equation 4.4 with actual numbers, we may use a genetic factor affecting coat coloration in guinea pigs (Table 4.2). The alleles at the locus are c^r and c^d, but for consistency with other symbols in this chapter we will designate them as A and A'. The phenotypic value of each animal is measured as $\arcsin\sqrt{x}$, where x is the proportion of black coloration on the animal. The mean phenotypes of AA, AA', and $A'A'$ genotypes are denoted a, d, and $-a$, respectively, which are measured as a deviation from the average of the homozygous genotypes (in this case 61.60). The calculations of a and d are shown in column 3. The symbols a and d represent the effects of the alleles. The quantity $2a$ measures the difference between means of the homozygous genotypes, because $a - (-a) = 2a$, and d/a serves as a measure of dominance:

- $d = a$ means that A is dominant to A'
- $d = 0$ implies additivity (AA' exactly intermediate between AA and $A'A'$)
- $d = -a$ means that A' is dominant to A

In this example $a = 7.27$ and $d = -0.93$. The negative sign on d means that the A' allele (c^d) is partially dominant, and $d/a = -0.128$.

To obtain the G_i values needed in Equation 4.4, we need to express the mean phenotype of each genotype as a deviation from the overall population mean. If A and A' have allele frequencies p and q, then in a random mating population the overall mean phenotypic value μ is given by $p^2a + 2pqd + q^2(-a)$. The deviations are obtained by subtracting μ from the mean of each genotype as shown in column 4. The algebra is somewhat tedious, but each deviation can be expressed in terms of the quantity $a + (q-p)d$, which appears so often in the equations of quantitative genetics that it is assigned the special symbol α and given the special name **average effect**

$$\alpha = a + (q - p)d \qquad (4.6)$$

The average effect can also be written as $q[d - (-a)] + p(a - d)$, which more easily shows its biological meaning. It is the average change in mean phenotype that would result from choosing an A' allele at random (in whatever genotype it happens to be) and changing it into an A allele.

Components of Genotypic Variation

The genotypic variance σ_g^2 due to the locus in Table 4.2 is therefore

$$\sigma_g^2 = p^2\big[2q\alpha - 2q^2d\big]^2 + 2pq\big[(q-p)\alpha + 2pqd\big]^2 + q^2\big[-2p\alpha - 2p^2d\big]^2 \qquad (4.7)$$

Table 4.2 Expressions for allelic effects of one locus affecting coat coloration in guinea pigs

Genotype	Mean phenotype value	Deviation from average of the homozygous genotypes	Deviation from population mean (allele frequencies p and q)
$c^r c^r (AA)$	68.87	$a = 68.87 - 61.60 = 7.27$	$G_1 = a - \mu$ $= -2q[a + (q-p)d] - 2q^2 d$ $= 2q\alpha - 2q^2 d$
$c^r c^d (AA')$	60.67	$d = 60.67 - 61.60 = -0.93$	$G_2 = d - \mu$ $= (q-p)[a + (q-p)d] + 2pqd$ $= (q-p)\alpha + 2pqd$
$c^d c^d (A'A')$	54.33	$-a = 54.33 - 61.60 = -7.27$	$G_3 = -a - \mu$ $= -2p[a + (q-p)d] - 2p^2 d$ $= -2p\alpha - 2p^2 d$

$$\text{Population mean } \mu = p^2 a + 2pqd + q^2(-a)$$
$$= (p - q)a + 2pqd$$

Source: Data from Wright (1968).

Here again the algebra is rather tedious, but it helps to consider the coefficients of the α^2, αd, and d^2 terms separately. These are:

$$\left[4p^2 q^2 + 2pq(q-p)^2 + 4p^2 q^2\right] * \alpha^2 = 2pq * \alpha^2$$
$$\left[-8p^2 q^3 + 8p^2 q^2(q-p) + 8p^3 q^2\right] * \alpha d = 0$$
$$\left[4p^2 q^4 + 8p^3 q^3 + 4p^4 q^2\right] * d^2 = (2pq)^2 d^2$$

Hence, we can write

$$\sigma_g^2 = 2pq\alpha^2 + (2pqd)^2 \tag{4.8}$$

The first term in Equation 4.8 is called the **additive genetic variance,** symbolized σ_a^2:

$$\sigma_a^2 = 2pq\alpha^2 \tag{4.9}$$

The second term is called the **dominance variance,** symbolized σ_d^2:

$$\sigma_d^2 = (2pqd)^2 \tag{4.10}$$

Whereas the additive variance depends both on a (twice the difference between homozygous genotypes) and d (the dominance effect), the dominance variance depends only on d. Both variance components also depend on allele frequencies, as shown in Figure 4.5. The case of overdominance (Figure 4.5D) shows that the additive genetic variance can equal 0 (in this example at $p = q = 1/2$), even though the genotypic variance is nonzero. It is a worthwhile exercise to show that the additive genetic variance equals $8pq^3a^2$ for A dominant and $8p^3qa^2$ for A recessive.

When multiple loci are considered, $2pq\alpha^2$ in Equation 4.9 is replaced with the sum of such terms, one for each locus, and likewise $(2pqd)^2$ in Equation

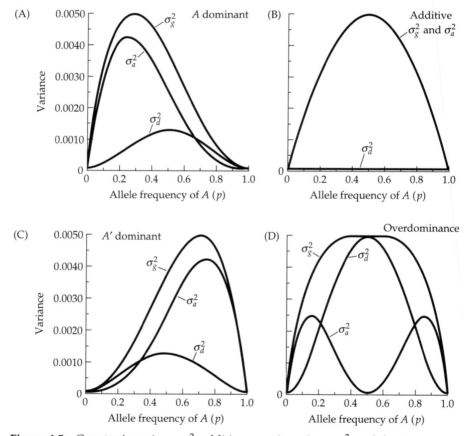

Figure 4.5 Genotypic variance σ_g^2, additive genetic variance σ_a^2, and dominance variance σ_d^2 for a gene with two alleles A and A' at allele frequencies p and q in a population in HWE. The mean phenotypes of AA, AA', and $A'A'$ are a, d, and $-a$, respectively, with values chosen to maximize σ_g^2 at 0.005. (A) $a = d = 0.07071$. (B) $a = 0.1$, $d = 0$. (C) $-a = d = 0.07071$. (D) $a = 0$, $d = 0.14142$.

4.10 is replaced with a sum of similar terms. Each locus may have different values of p, q, a, and d. There is also a term corresponding to nonadditive interactions between the genotypes at different loci, which is called the **interaction variance,** symbolized σ_i^2. And in still more general models in which assortative mating is allowed, there is a term due to assortative mating, symbolized σ_{am}^2. (The term **assortative mating** means that there is a positive correlation between the phenotypes of mating pairs, as occurs in human populations for height.) With all these complications taken into account, the genotypic variance can be written as

$$\sigma_g^2 = \sigma_a^2 + \sigma_d^2 + \sigma_i^2 + \sigma_{am}^2 \tag{4.11}$$

In principle, one could also partition the environmental variance. For example, with respect to throat cancer, there would be an environmental effect due to cigarette smoking ($\sigma_{smoking}^2$), one due to alcohol consumption ($\sigma_{drinking}^2$), and one due to the synergistic effects of smoking and drinking ($\sigma_{both\ together}^2$). For these two environmental factors we could then write

$$\sigma_e^2 = \sigma_{smoking}^2 + \sigma_{drinking}^2 + \sigma_{both\ together}^2 \tag{4.12}$$

However, the analysis of environmental causes of variation in complex traits is more difficult than that of genetic causes. While the elementary genetic factors are genes, it is often very difficult to identify critical environmental factors.

ARTIFICIAL SELECTION

It should be clear intuitively that selection cannot change the genotype of a population in which every individual has an identical genotype, such as a highly inbred line. Selection is ineffective in such cases because the only causes of variation are environmental, which are not transmitted from generation to generation. Genetic changes can sometimes occur slowly in traits affected by many genes in populations that are large enough, because then selection can act on the genetic variation contributed by new mutations. But for bristle number in inbred strains of *Drosophila*, new mutations arising in each generation account for only 0.1–1% of the variance in bristle number (Mackay et al. 1994).

Although genetic variation is essential for progress under selection, it is not sufficient. The reason is really quite subtle. Only the additive genetic variance σ_a^2 contributes to the response to selection. The genotypic variance σ_g^2 is not the key quantity. We will examine this principle by examining **artificial selection,** which refers to the deliberate choice of a select group of individuals to be used for breeding. The most common type of artificial selection is **directional selection,** in which phenotypically superior individuals are chosen. Artificial selection has been practiced empirically for thousands of years, for example, in the body size of domesticated dogs (Vilá et al. 1997). But understanding the genetic principles permits the breeder to predict the speed and degree to which a population can be changed through artificial selection in any small number of generations. Figure 4.6 shows the results of long-term artificial selection for oil content in corn. The "up" line illustrates the important point that, after many generations of directional selection, the *mean* phenotypic value of the selected population can greatly exceed the *maximum* phenotypic value among individuals in the original population. Not as much progress was made in the "down" line, because 0% oil is an absolute lower limit.

In most genetically heterogeneous populations, artificial selection can change phenotype well beyond the range of variation found in the original population. After 100 generations of selection for pupa weight in the flour beetle *Tribolium castaneum,* the selected population had a mean pupa weight about 17 phenotypic standard deviations greater than the mean of the origi-

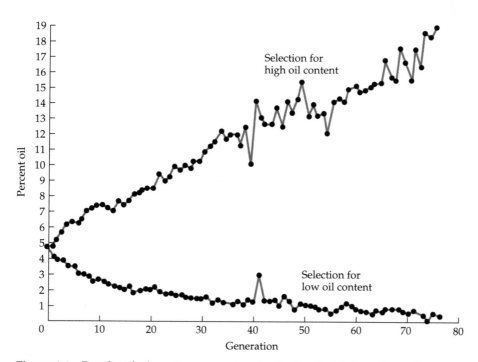

Figure 4.6 Results of a long-term experiment selecting for high and low oil content in corn seeds. Begun in 1896, the experiment has the longest duration of any on record and still continues at the University of Illinois. (After Dudley 1977.)

nal population (Enfield 1980). This is an exceptional example, but a total selection response of 3–5 times the original phenotypic standard deviation is not unusual, and for selection to change a population of effective size N_e halfway to its selection limit typically requires about $0.5N_e$ generations (Falconer 1977).

Prediction Equation for Individual Selection

When each individual is selected for breeding based solely on its own phenotypic value, the type of artificial selection is called **individual selection**. Figure 4.7 illustrates a type of individual selection called **truncation selec-**

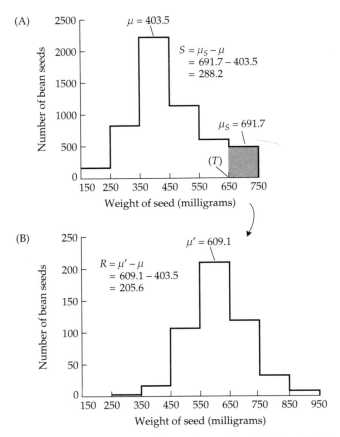

Figure 4.7 Truncation selection for seed weight in edible beans of the genus *Phaseolus*. The truncation point is $T = 650$ mg. (Data from Johannsen 1903.)

tion. The trait is seed weight in edible beans, and the experiment is one of the first of its kind (Johannsen 1903). The histogram in (A) represents the distribution of seed weight in the parental population. The shaded part of the distribution to the right of the phenotypic value denoted T (650 mg) indicates those seeds selected to germinate and grow for breeding among themselves. The value T is called the **truncation point.** The mean phenotype in the entire population is denoted μ (403.5 mg), and that of the selected parents is denoted μ_S (691.7 mg). When the selected parents were mated at random, their offspring seeds had the phenotypic distribution shown in (B), where the mean phenotype is denoted μ' (609.1 mg). It is typical of truncation selection that the offspring mean μ' is greater than μ but less than μ_S. The reason μ' is greater than μ is that some of the selected parents have favorable genotypes and so transmit favorable genes to their offspring. At the same time, μ' is generally less than μ_S for two reasons:

- Some of the selected parents do not have favorable genotypes. Their exceptional phenotypes result from chance exposure to exceptionally favorable environments.
- Alleles, not genotypes, are transmitted from parents to offspring, and the favorable genotypes in the parents are disrupted by Mendelian segregation and recombination.

The difference in mean phenotype between the selected parents and the entire parental population is called the **selection differential** and designated S:

$$S = \mu_S - \mu \qquad (4.13)$$

The difference in mean phenotype between the progeny generation and the previous generation is called the **response to selection** and designated R:

$$R = \mu' - \mu \qquad (4.14)$$

For the data in Figure 4.7, $S = 288.2$ mg and $R = 205.6$ mg.

Any equation that defines the relationship between the selection differential S and the response to selection R is known as a **prediction equation.** Each type of artificial selection has its own prediction equation. For individual truncation selection, the prediction equation is

$$R = h^2 S \qquad (4.15)$$

where h^2 is called the **narrow-sense heritability** of the trait. As with the broad-sense heritability in Equation 4.5, the narrow-sense heritability is a ratio of variances, but not the ratio of genotypic variance to phenotypic variance. Rather, as we shall see, the narrow-sense heritability is the ratio of the additive genetic variance to the phenotypic variance,

$$h^2 = \frac{\sigma_a^2}{\sigma_p^2} \tag{4.16}$$

But for the moment let us interpret the narrow-sense heritability merely as a description of an observed result. In Figure 4.7, for example, $S = 288.2$ mg and $R = 205.6$; hence, $h^2 = R/S = 205.6$ mg$/288.2$ mg $= 71.3\%$. When estimated from an observed result, h^2 is called the **realized heritability.**

The usual way to estimate the narrow-sense heritability is from the correlation between relatives. To establish the basis for this approach we must consider the genetics behind Equation 4.15. This entails examining how alternative alleles of a gene affect a complex trait, how truncation selection changes the allele frequencies, and how much any change in allele frequency changes the mean phenotypic value.

Genetic Basis of Complex Traits

Figure 4.8 shows a normal distribution of a complex trait in a hypothetical randomly mating population. The mean is μ and the phenotypic variance σ_p^2. In truncation selection, all individuals with phenotypes above the truncation point T are saved for breeding, and the shaded area B of the distribution rep-

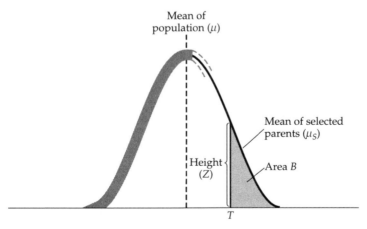

Figure 4.8 Normal distribution of a quantitative trait in a population subjected to truncation selection with truncation point T.

resents the proportion of the population selected. The height of the normal distribution at the point T is denoted Z, so $Z = f(T)$ in Equation 4.2. The mean phenotype among the selected individuals is μ_S. One of the special properties of the normal distribution to be used later is that

$$\frac{\mu_S - \mu}{\sigma_p^2} = \frac{Z}{B} \tag{4.17}$$

To determine the increase in mean phenotype of a population from one generation of truncation selection, we first imagine a gene that affects the trait with alleles A and A' at respective allele frequencies p and q. Because of random mating, genotypes AA, AA', and $A'A'$ are present in the population with frequencies p^2, $2pq$, and q^2, respectively, but the individual genotypes cannot be identified through their phenotypic values because of the variation in phenotype caused by environmental factors and genetic differences at other loci. If the genotypes could be identified, their distributions of phenotypic value would have slightly different means, as shown in Figure 4.9. As in Table 4.2, the mean phenotypes of AA, AA', and $A'A'$ genotypes are denoted

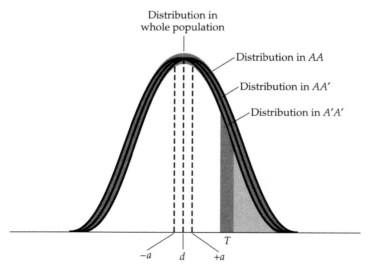

Figure 4.9 Same distribution as in Figure 4.8, showing the slightly displaced distributions of phenotypic value among the three genotypes AA, AA', and $A'A'$ for a gene that contributes to the quantitative trait. The means of the distributions are a, d, and $-a$, respectively.

a, d, and $-a$, respectively. The mean phenotype of a population in Hardy-Weinberg equilibrium for A and A' is therefore given by

$$\mu = p^2 a + 2pqd + q^2(-a) = (p-q)a + 2pqd \qquad (4.18)$$

Next we need to calculate the change in allele frequency that takes place as a result of selection.

Change in Allele Frequency

Suppose for the moment that we were practicing artificial selection for increased amount of black coat coloration in the guinea pigs in Table 4.2. Selection for black coat coloration in a population containing both the c^r (A) and c^d (A') alleles would be successful in increasing the allele frequency of A, and the average amount of black coloration among animals of the next generation would increase. To calculate the expected increase in black coloration in one generation of selection, we must first calculate the change in the allele frequency of A. An equation for change in allele frequency with natural selection was derived in Chapter 2 (Equation 2.31), which remains valid for artificial selection if we agree to interpret the "fitness" of an individual as the probability that the individual is included among the group selected as parents of the next generation. With this interpretation of fitness, differences in fitness (reproductive success) of AA, AA', and $A'A'$ genotypes correspond to the differences in area to the right of the truncation point in Figure 4.9, because only those individuals in the shaded area are selected to reproduce. The differences in area are easy to calculate if you shift or slide each curve horizontally until the means coincide. Shift the $A'A'$ curve a units to the right, and shift the AA' and AA curves d and a units to the left. This brings the distributions into coincidence, but it slides the truncation points slightly out of register, as shown in Figure 4.10. The difference in "fitness" between AA and AA', denoted $w_{AA} - w_{AA'}$, is equal to the small area indicated in Figure 4.10, as is the difference in fitness between AA' and $A'A'$, denoted $w_{AA'} - w_{A'A'}$. The areas corresponding to $w_{AA} - w_{AA'}$ and $w_{AA'} - w_{A'A'}$ are approximately rectangles, and the area of a rectangle is the product of the base and the height. Therefore, since Z represents the height of the normal distribution at the point T, we can make the following approximations:

$$
\begin{aligned}
w_{AA} - w_{AA'} &= Z\big[(T+a)-(T+d)\big] = Z(a-d) \\
w_{AA'} - w_{A'A'} &= Z\big[(T+d)-(T-a)\big] = Z(a+d)
\end{aligned}
\qquad (4.19)
$$

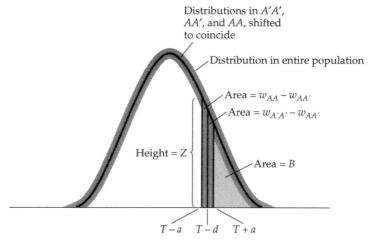

Figure 4.10 Same distribution as in Figures 4.8 and 4.9, but with the distributions of AA, AA', and $A'A'$ shifted to coincide. Shifting the distributions slides the truncation points slightly out of register, so the truncation points become $T - a$, $T - d$, and $T + d$, respectively.

The average fitness $\bar{w}$ of the entire population simply equals B, because B is the proportion of the population saved for breeding. From Equation 2.31 we know that $\Delta p = pq[p(w_{AA} - w_{AA'}) + q(w_{AA'} - w_{A'A'})]/\bar{w}$, where Δp is the change in frequency of the allele A in one generation of selection. Substituting from Equation 4.19 and using $\bar{w} = B$ leads to

$$\Delta p = \frac{pq[pZ(a-d) + qZ(a+d)]}{B}$$

or, since $p + q = 1$,

$$\Delta p = \frac{Z}{B} pq[a + (q-p)d] \tag{4.20}$$

An equation corresponding to 4.20 could be obtained for any gene affecting the trait, but the values of p, q, a, and d would differ for each gene.

Change in Mean Phenotype

Equation 4.20 provides an expression for Δp that can be used to calculate the mean phenotypic value in the next generation. Among the progeny of selected parents the allele frequencies of A and A' are $p + \Delta p$ and $q - \Delta p$,

respectively. With random mating, the mean phenotype in the next generation is given by Equation 4.18 as

$$\mu' = (p + \Delta p)^2 a + 2(p + \Delta p)(q - \Delta p)d + (q - \Delta p)^2(-a)$$

When the right-hand side of this expression is multiplied out and terms in $(\Delta p)^2$
are ignored because Δp is usually small, then μ' is found to be approximately

$$\mu' = \mu + 2[a + (q - p)d]\Delta p \tag{4.21}$$

Now move μ to the left-hand side, substitute Δp from Equation 4.20, and
replace Z/B with the left-hand side of Equation 4.17. The result is

$$\mu' - \mu = (\mu_S - \mu)\frac{2pq[a + (q - p)d]^2}{\sigma_p^2} \tag{4.22}$$

But $S = \mu_S - \mu$ is the selection differential (Equation 4.13), $R = \mu' - \mu$ is the
response to selection (Equation 4.14), and $h^2 = R/S$ (Equation 4.15). Hence,
Equation 4.22 implies that

$$h^2 = \frac{2pq[a + (q - p)d]^2}{\sigma_p^2} = \frac{2pq\alpha^2}{\sigma_p^2} = \frac{\sigma_a^2}{\sigma_p^2} \tag{4.23}$$

This is the equation that we wanted, because it defines the narrow-sense heritability in terms of its variance components, and justifies Equation 4.16. We
have thereby shown that:

- The narrow-sense heritability h^2 equals the ratio of the additive genetic
  variance to the phenotypic variance (Equation 4.23).
- The narrow-sense heritability h^2 is the heritability used in the prediction
  equation for individual selection (Equation 4.15).

Application of these equations can be illustrated using the genetic difference for coat color in Table 4.2, where $a = 7.27$ and $d = -0.93$. Consider a random-mating population in which $p = q = 1/2$. With equal allele frequencies
the term in d in Equation 4.23 disappears, and then $\sigma_a^2 = 2(1/2)(1/2)a^2 = 26.43$.
This is close to its maximum possible value for $a = 7.27$ and $d = -0.93$, which
occurs when $p = 0.562$ yielding $\sigma_a^2 = 26.852$. The narrow-sense heritability due
to this locus depends on the phenotypic variance σ_p^2, but h^2 is minimized at
$p = 0$ or $p = 1$ (for which $h^2 = 0$) and maximized at $p = 0.562$.

Equation 4.23 is valid only when a single gene affects the trait, but for multiple genes the right-hand side is replaced by a summation of such terms, one for each gene. The equivalent equation for multiple genes is therefore

$$h^2 = \frac{\sum_i 2p_i q_i \left[a_i + (q_i - p_i)d_i\right]^2}{\sigma_p^2} = \frac{2pq\alpha^2}{\sigma_p^2} = \frac{\sigma_a^2}{\sigma_p^2} \qquad (4.24)$$

where p_i, q_i, a_i, and d_i are the values of p, q, a, and d for the ith gene, and the summation is over all genes affecting the complex trait. The summation in the numerator is the additive genetic variance of the trait due to all loci.

CORRELATION BETWEEN RELATIVES

Estimation of the additive genetic variance might at first seem to be very difficult, but in fact it is quite straightforward. The reason is that the theoretical covariance between certain types of relatives is a simple multiple of σ_a^2. Suppose x and y represent the phenotypic values of a trait between any pair of relatives, for example, first cousins. Then the **covariance** of x and y is defined as

$$\text{Cov}(x,y) = E[x - E(x)][y - E(y)] \qquad (4.25)$$

Just as the variance is the expected squared deviation from the mean, the covariance is the expected product of deviations from the mean. If x and y are independent, then $\text{Cov}(x, y) = 0$. Based on a random sample of n pairs of values (x_i, y_i), the covariance is estimated as

$$\langle \text{Cov}(x,y) \rangle = \frac{\sum_{i=1}^{n} (x_i - \bar{x})(y_i - \bar{y})}{n - 1} \qquad (4.26)$$

The covariance is used in calculating the **correlation coefficient** r in phenotypic value between x and y:

$$r = \frac{\text{Cov}(x,y)}{\sigma_x \sigma_y} \qquad (4.27)$$

where σ_x and σ_y are the phenotypic standard deviations of x and y. For the examples that we will consider, $\sigma_x = \sigma_y = \sigma_p$, the phenotypic standard deviation in the population. The correlation coefficient is sometimes preferred over the covariance as a measure of relationship between random variables, because r must always lie between -1 (perfect negative correlation, $x = -y$) and $+1$ (perfect positive correlation, $x = y$). Independence between x and y implies that $r = 0$.

Parent-Offspring Correlation

The parent-offspring correlation is often used to estimate the narrow-sense heritability because the covariance equals the additive genetic variance. The covariance can be calculated either from a single parent (usually chosen to be the father, to avoid possible nongenetic maternal effects on the offspring) or from the average of the parents (called the **midparent**). The result is the same. The calculations will be illustrated using father-offspring pairs in a random mating population, as set out in Table 4.3. The first two columns give each parental genotype with its frequency and the mean phenotypic deviation. The next two columns show the gametes produced by the parent and the frequencies of offspring genotypes with random mating. The two columns at the right give the frequencies of parent-offspring pairs and the mean phenotypic deviation of the offspring. The covariance is calculated as the product of the parent-offspring deviations, each weighed by the frequency of the parent-offspring pair. The first term, for example, is $p^3(2q\alpha - 2q^2d)^2$. In the sum of these products, the coefficient of the term α^2 is:

$$4p^3q^2 + 4p^2q^2(q-p) + pq^2(q-p)^2 + p^2q(q-p)^2 - 4p^2q^2(q-p) + 4p^2q^3 = pq$$

Table 4.3 Frequencies and phenotypic deviations of parent-offspring pairs with random mating

Parental genotype (frequency)	Parental deviation	Gametes	Uniting gamete (frequency)	Offspring genotype (frequency)	Offspring deviation
AA (p^2)	$2q\alpha - 2q^2d$	A	A (p)	AA (p^3)	$2q\alpha - 2q^2d$
			A' (q)	AA' (p^2q)	$(q-p)\alpha + 2pqd$
AA' $(2pq)$	$(q-p)\alpha + 2pqd$	$\frac{1}{2}$ A	A (p)	AA (p^2q)	$2q\alpha - 2q^2d$
			A' (q)	AA' (pq^2)	$(q-p)\alpha + 2pqd$
		$\frac{1}{2}$ A'	A (p)	AA' (p^2q)	$(q-p)\alpha + 2pqd$
			A' (q)	$A'A'$ (pq^2)	$-2p\alpha - 2p^2d$
$A'A'$ (q^2)	$-2p\alpha - 2p^2d$	A'	A (p)	AA' (pq^2)	$(q-p)\alpha + 2pqd$
			A' (q)	$A'A'$ (q^3)	$-2p\alpha - 2p^2d$

As an exercise, one may verify that the coefficients of the αd and the d^2 terms are both 0. Therefore, letting Cov(PO) be the covariance between a single parent and its offspring,

$$\text{Cov}(PO) = pq\alpha^2 = \frac{\sigma_a^2}{2} \tag{4.28}$$

Alternatively we can write the correlation coefficient between a single parent and its offspring r_{PO} as

$$r_{PO} = \frac{\text{Cov}(PO)}{\sigma_p^2} = \frac{(1/2)\sigma_a^2}{\sigma_p^2} = \frac{h^2}{2} \tag{4.29}$$

where h^2 is the narrow-sense heritability. Either Equation 4.28 or 4.29 may be used as a basis for estimating h^2.

Heritability Estimates from Resemblance between Relatives

Theoretical covariances for several common relationships are shown in Table 4.4. The additive genetic variance can be estimated directly from covariance between parent and offspring, midparent (the average of the parents) and offspring, half siblings, uncle-nephew (or aunt-niece), or first cousins. What these degrees of relationship have in common is that the relatives can share at most one allele at any locus. The covariances between the other degrees of

Table 4.4 Theoretical covariance in phenotype between relatives[a]

Degree of relationship	Covariance
Offspring and one parent	$\sigma_a^2/2$
Offspring and average of parents (midparent)	$\sigma_a^2/2$
Half siblings	$\sigma_a^2/4$
Full siblings	$(\sigma_a^2/2) + (\sigma_d^2/4)$
Monozygotic twins	$\sigma_a^2 + \sigma_d^2$
Nephew and uncle	$\sigma_a^2/4$
First cousins[b]	$\sigma_a^2/8$
Double first cousins[b]	$(\sigma_a^2/4) + (\sigma_d^2/16)$

[a]Variance terms due to interaction between loci (epistasis) are ignored.
[b]First cousins are the offspring of matings between siblings and unrelated individuals; double first cousins are the offspring of matings between siblings from two different families.

relationship in Table 4.4 include a contribution due to dominance because the relatives can share two alleles at any locus. The theoretical covariance between monozygotic twins is equal to the covariance of an individual with itself, or σ_g^2 (Equation 4.8).

Figure 4.11 shows estimates of narrow-sense heritabilities of diverse quantitative traits as estimated from the correlation between relatives. The data are presented merely to show the values of heritability that breeders typically consider. The heritabilities pertain to one population at one particular time. The same trait in a different population or in a different environment might well have a different heritability. Generally speaking, traits that are closely related to fitness (such as calving interval in cattle or eggs per hen in poultry) tend to have rather low heritabilities. For comparison, Figure 4.11 also shows estimated broad-sense heritabilities of a number of quantitative

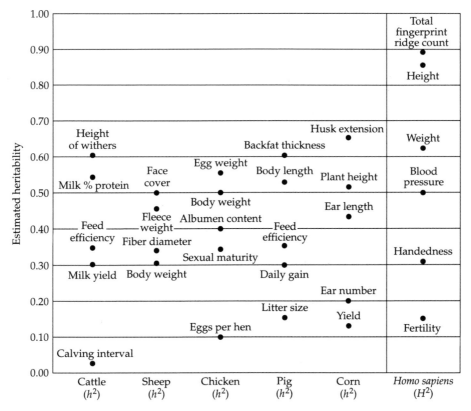

Figure 4.11 Narrow-sense heritabilities of representative traits in domesticated animals and plants, and broad-sense heritabilities of various traits in human beings. (Data from Pirchner 1969; Robinson et al. 1949; and Smith 1975.)

traits in human populations. Broad-sense heritabilities vary widely for different traits, as they do in other species. Note the relatively low heritability of fertility, a trait that is obviously closely related to fitness.

Offspring-on-Parent Regression

Figure 4.12 shows a plot of parental phenotype along the x-axis and offspring phenotype along the y-axis. The straight line is the **regression line** of offspring on parent, determined by finding the slope and intercept that minimizes the sum of the squares of the vertical deviation of the points from the line. Parent-offspring regression is a convenient method for estimating the narrow-sense heritability because the slope of the line b_{OP}, called the **regression coefficient,** can be shown to satisfy

$$b_{OP} = \frac{\text{Cov}(PO)}{\sigma_p^2} = \frac{(1/2)\sigma_a^2}{\sigma_p^2} = \frac{h^2}{2} \tag{4.30}$$

where $\text{Cov}(PO)$ is the covariance of one parent and its offspring and σ_p^2 is the phenotypic variance among the parents (as well as that among the offspring). Hence, h^2 can be estimated as $2b_{OP}$. (The regression coefficient of midparent on offspring, b_{OM}, equals h^2.) It makes no difference whether the offspring of each parent are considered individually or are pooled and their mean used instead. Figure 4.12 uses the mean pupa weight of each set of progeny. The

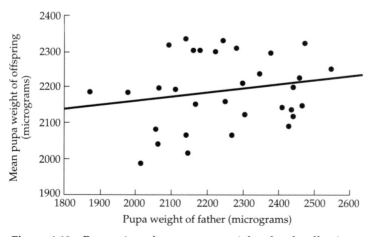

Figure 4.12 Regression of mean pupa weight of male offspring on father's pupa weight in the flour beetle *Tribolium castaneum*. Each point is the mean of about eight male offspring. (Data courtesy of F. D. Enfield.)

slope of the regression line is $b_{OP} = 0.11$, and thus the estimate of the narrow-sense heritability of pupa weight in this population is $h^2 = 2b_{OP} = 0.22$. The data points represent 32 sibships and are quite scattered. Because of this sort of scatter, heritability estimates tend to be rather imprecise unless based on data from several hundred families.

QUANTITATIVE GENETICS OF NATURAL POPULATIONS

There is a large literature dealing with issues related to the genetic components of phenotypic variation in natural populations and to the long-term evolution of quantitative traits. The main issues are reviewed in Caballero and Keightley (1994), Bürger and Lande (1994), and Turelli and Barton (1994). The underlying problem is how to explain the genetic differences that exist among quantitative traits in natural populations. The kinds of differences are illustrated in Table 4.5, which summarize some of the important genetic components of several quantitative traits in *Drosophila* (Houle 1998). The symbol σ_m^2 is the **mutational variance,** defined as the increment of variance generated each generation by mutation assuming additive effects on a metric trait and selective neutrality (Lynch and Hill 1986; Lynch 1988). The symbol σ_{TEI}^2 is the estimated average amount of variation produced by a new mutation resulting from the insertion of a single transposable element. The additive genetic variance is denoted σ_a^2. Each of these sources of variation is compared as the **coefficient of variation,** defined as the standard deviation expressed as a percentage of the mean ($100\sigma/\bar{x}$), which enables comparison of traits measured on different scales. The quantity σ_a^2/σ_m^2 is the additive genetic variance

Table 4.5 Genetic variation and fitness sensitivity of traits in *Drosophila melanogaster*

Trait	Mutation $100\dfrac{\sqrt{\sigma_m^2}}{\bar{x}}$	Single insertion $100\dfrac{\sqrt{\sigma_{TEI}^2}}{\bar{x}}$	Additive effects $100\dfrac{\sqrt{\sigma_a^2}}{\bar{x}}$	Additive/ mutation $\dfrac{\sigma_a^2}{\sigma_m^2}$	Fitness sensitivity $\dfrac{\Delta\bar{w}}{1\%\Delta\bar{x}}$
Abdominal bristles	0.24	2.04	6.11	646.01	0.03
Sternopleural bristles	0.39	3.01	7.38	367.97	0.04
Developmental time	0.43	1.20	2.47	33.72	1.00
Viability	1.57	38.50	10.40	43.75	1.00
Early fecundity	1.22	19.74	8.81	52.12	0.96
Late fecundity	2.56	50.46	28.79	126.23	0.04
Longevity	1.35	14.44	9.06	45.22	0.00

Source: Data from Houle (1998).

expressed as a multiple of the mutational variance, and the "fitness sensitivity" is an estimate of how closely each metric trait is tied to fitness as measured by the percentage decrease in fitness expected for a 1% change in the mean of the metric trait in the unfavorable direction. The calculation assumes that each metric trait affects fitness as a whole. There are no tradeoffs between fitness components, such as when two or more components of fitness are affected in opposing directions. The finding that genes may affect several traits simultaneously is called **pleiotropy**.

The differences among the traits in Table 4.5 are manifest. The bristle traits are affected less by new mutations than other traits, the additive variance is a large multiple of the mutational variance, and the fitness sensitivity is quite small. These traits seem to fit a model in which the mutational variance contributed in each generation accumulates over time. For a metric trait affected by mutations that are additive in their effects and selectively neutral, the additive genetic variance is expected to change in each generation according to

$$\sigma_a^2(t+1) \approx \left(1 - \frac{1}{2N_e}\right)\sigma_a^2(t) + \sigma_m^2$$

where N_e is the effective population number (Clayton and Robertson 1955; Lynch and Hill 1986). Hence at equilibrium

$$\sigma_a^2(t+1) = \sigma_a^2(t) = 2N_e\sigma_m^2 \qquad (4.31)$$

For *Drosophila*, N_e is on the order of 10^6 and σ_m^2 is on the order of 10^{-3}–10^{-4} σ_a^2, hence values of σ_a^2/σ_m^2 in the range 100–1000 might be expected from mutation accumulation alone, especially considering that Equation 4.31 is an overestimate when the assumption of selective neutrality is relaxed.

As might be expected intuitively, for traits that are strongly affected by new mutations and that have a high fitness sensitivity (developmental time, viability, and early fecundity), σ_a^2/σ_m^2 is reduced by an order of magnitude. What about late fecundity and longevity? They have a low fitness sensitivity yet a relatively small ratio of σ_a^2/σ_m^2. These traits are probably strongly affected by pleiotropic effects of genes affecting developmental time and early fecundity, and therefore the low fitness sensitivities are probably misleading because the genes are subjected to selection at an earlier stage in life (Houle 1998).

Directional Selection with Mutation-Selection-Drift

Theoretical models of the evolution of quantitative traits tend to be quite complex because their parameters include the number of loci affecting the trait, mutation rates, allele frequencies, distribution of effects of alleles, dominance relationships between the alleles, interactions between alleles at different loci, pleiotropic effects on different traits, and recombination rates (Barton and Turelli 1989; Turelli and Barton 1990). Many of these parameters are unknown, and estimates that are available come from a handful of traits primarily in *Drosophila*. Nevertheless, taking what information is available into account, Caballero and Keightley (1994) have analyzed a model of mutation-selection-drift for a metric trait that incorporates:

- Mutational effects that are more concentrated around the mean than a normal distribution (leptokurtic)
- Negative pleiotropic effects on fitness with an intermediate correlation between the metric trait and fitness
- Near recessivity of mutant alleles of large effect and near additive effects of mutant alleles of small effect

In the Caballero-Keightley model, the equilibrium genotypic variance maintained for bristle number is in good agreement with observations, assuming that a relatively small number of genes affects the trait. For population size $N = 10^4$–10^6, the equilibrium genotypic variance σ_g^2 is nearly independent of the degree of dominance of new mutations, with the dominance variance $\sigma_d^2 \approx 0.10\,\sigma_g^2$ and the narrow-sense heritability $h^2 \approx 0.4$–0.6. Most of the equilibrium genotypic variance is due to about 1% of the mutations that happen to have very small effects on fitness and intermediate effects on the trait ($a = 0.1$–$0.5\,\sigma_p$). Consistent with the low values of narrow-sense heritability of traits strongly correlated with fitness (Figure 4.11), if the metric trait is assumed to be viability and all new mutations affect the trait, then $h^2 \approx 0.1$ independent of population size.

Stabilizing Selection with Mutation-Selection-Drift

Some traits are thought to be subjected to **stabilizing selection,** in which selection favors individuals with phenotypes near the population mean. Deviation from the mean in either direction is deleterious. A classic example is birth weight in human babies prior to the advent of postnatal intensive care (Karn and Penrose 1951), when the probability of survival decreased with the deviation of the birth weight from the mean (Figure 4.13). Since the fitness sensitivity of viability is very high, in this case the connection between the metric trait and fitness is causal.

In a mutation-selection-drift model of stabilizing selection, in which the fitness of each phenotype falls off as the square of the deviation from the mean, for realistic values of the parameters the equilibrium σ_g^2/σ_m^2 falls in the range of values in Table 4.5 (Bürger and Lande 1994). The main difference from the pleiotropic model discussed in the previous section is that, when stabilizing

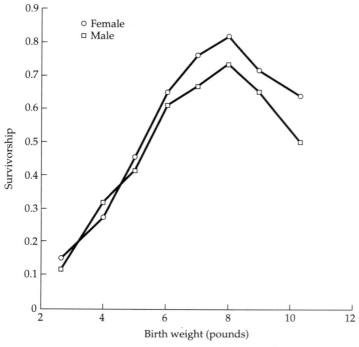

Figure 4.13 Stabilizing selection for birth weight in human babies prior to modern neonatal intensive care. (Data from Karn and Penrose 1951.)

selection is strong, the equilibrium genotypic variance is low; and when the strength of stabilizing selection is weak, most of the equilibrium genotypic variance is due to the segregation of very deleterious alleles ($Ns < -5$), and mutations of small effect on the metric trait ($a < 0.125\sigma_p$) contribute a disproportionate share to the genotypic variance (Caballero and Keightley (1994).

Antagonistic Pleiotropy

When genotypes indirectly affect components of fitness in opposite directions, the situation is called **antagonistic pleiotropy.** For example, directional selection for rapid larval development in *Drosophila* leads to a reduction in adult body weight (Nunney 1996). This is known as a **correlated response.** Just as the additive part of the genotypic variance for one trait (Equation 4.9) determines the correlation between relatives (Equation 4.29) and the response to selection (Equation 4.23), the additive part of the covariance between traits determines the **genetic correlation** between relatives and the correlated response to selection (Falconer and Mackay 1996). Genetic correlations can result from interactions between metabolic, physiological, or developmental pathways or from genetic linkage, and they can sometimes be diminished or eliminated over the course of long-term selection (Leroi et al. 1994). Antagonistic pleiotropy does not seem to be a common feature of fitness-related traits in natural populations (Draye et al. 1994). In any case, the conditions for stable polymorphism are rather restrictive, especially with multiple loci and weak selection (Curtsinger et al. 1994). Furthermore, the equilibrium genetic variance is not as great as with directional or stabilizing selection, and the predicted dominance variance is greater than that observed. The conclusion from both experimental and theoretical studies is that antagonistic pleiotropy probably accounts for only a small part of the genotypic variance of the components of fitness, even though tradeoffs among these components may be common (Curtsinger et al. 1994).

COMPLEX TRAITS WITH DISCRETE EXPRESSION

Discrete traits are expressed in an all-or-none fashion, but they may nevertheless have complex inheritance in the sense that pedigrees show no obvious pattern of Mendelian transmission. Even a major gene can escape detection if there is **incomplete penetrance,** which means that the affected phenotype is not always expressed, or if the affected phenotype can also result from other genes or from environmental causes. A trait that can result from two or more different genotypes is said to show **genetic heterogeneity.** When pedigrees of unrelated affected individuals are pooled, as they usually are in human genetics, then genetic heterogeneity makes the pattern of inheritance very complex because two or more major genes are treated as if they were one.

Threshold Traits: Genes as Risk Factors

Some discrete traits are truly multifactorial. There are no single genes that are critical risk factors in themselves. These traits are determined by multiple genetic and environmental factors that act collectively to determine the risk of the trait being expressed. They are *threshold traits,* mentioned earlier in this chapter. An example is pyloric stenosis, an obstruction of the opening at the lower end of the stomach. Pyloric stenosis is a threshold trait because, while each individual is either affected or not affected, the risk of being affected is transmitted in pedigrees as if it were a quantitative trait. Affected parents do not necessarily transmit the condition to their offspring, but they do transmit genes that increase the risk. The next section shows how the underlying risk toward a threshold trait such a pyloric stenosis can be analyzed as a quantitative trait.

Heritability of Liability

The basic idea behind the quantitative genetics of threshold traits is illustrated in Figure 4.14. The normal curve in (A) represents the (unobservable) distribution of a hypothetical **liability** (or risk) toward the threshold trait, measured on a scale such that the mean value is $\mu = 0$ and the variance is 1. This curve represents the parental generation, and any parent with a liability above the threshold T_P actually expresses the trait. The proportion of affected individuals in the parental generation is denoted B_P, and the mean liability among affected parents is denoted μ_P.

Among the offspring of matings in which one parent is affected, the distribution of liability is given in panel (B). It has the same variance as in (A), but the mean is shifted to the right, to the position μ'. We can estimate the realized **heritability of liability** using Equations 4.13–4.15 as

$$h^2 = \frac{R}{S} = \frac{\mu' - \mu}{\mu_S - \mu} = \frac{\mu'}{\mu_S} = \frac{\mu'}{\left(\dfrac{\mu_P}{2}\right)} = \frac{2\mu'}{\mu_P} \tag{4.32}$$

The third equality follows from the fact that liability is measured on a scale in which $\mu = 0$. The next follows from the fact that only one parent is affected, hence the mean of the parents, μ_S, when one is affected (mean μ_P) and the other not (mean $\mu = 0$) is given by $\mu_S = (\mu_P + 0)/2 = \mu_P/2$.

Both μ' and μ_P can be estimated from the proportion of affected individuals in the parental generation (B_P) and among the offspring when one parent is affected (B_O). For pyloric stenosis, the incidence in fathers and their sons is $B_P = 0.005$ and $B_O = 0.05$ (Carter 1961). B_P and T_P must satisfy the equation

$$B_P = \int_{T_P}^{\infty} N(0,1)dx \tag{4.33}$$

where $N(0, 1)$ is given in Equation 4.2 with $\mu = 0$ and $\sigma = 1$. Given any value of B_P the corresponding value of T_P can be found from tables of integrals of the normal distribution, from mathematical analysis programs such as Mathematica®, or even from the approximations given in the legend of Figure 4.2, bearing in mind that the probabilities in Figure 4.2 are for both tails and so $2B_P$ should be used instead of B_P. When $B_P = 0.005$, as for pyloric stenosis, $T_P = 2.58$.

The way to find μ' is illustrated in Figure 4.14C, where the offspring distribution has been shifted μ' units to the left to coincide with the parental distribution. To maintain the same proportion of affected offspring, B_O, the truncation point must also be shifted μ' units to the left, hence the displace-

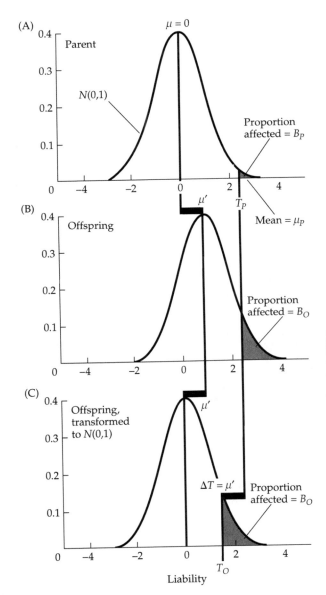

Figure 4.14 (A) Distribution of liability for a threshold trait, assumed to be normal with mean $\mu = 0$ and variance 1. The shaded area denotes parents with liability above a critical threshold (T_P) needed for expression of the trait. (B) Distribution of liability among offspring with one affected parent. The mean is displaced to μ'. (C) Shifting the offspring distribution to the left to coincide with $N(0,1)$ displaces the truncation point to $T_O = T_P - \mu'$.

ment in the truncation point is $\Delta T = T_P - T_O = \mu'$. However, the transformed distribution in panel (C) is $N(0,1)$ and so T_O satisfies

$$B_O = \int_{T_O}^{\infty} N(0,1)dx \tag{4.34}$$

and for $B_O = 0.05$ as in pyloric stenosis, $T_O = 1.64$. Therefore μ' in the numerator of Equation 4.32 is $\mu' = T_P - T_O = 2.58 - 1.64 = 0.94$.

The denominator in Equation 4.32 is the mean liability among the affected parents, which is given by

$$\mu_P = \frac{\int_{T_P}^{\infty} x N(0,1) dx}{\int_{T_P}^{\infty} N(0,1) dx}$$

This again requires evaluation using tables of the normal distribution or numerical integration, but in the case when $T_P = 2.58$, $\mu_P = 0.01430/0.00494 = 2.89$.

Putting all this together and substituting into Equation 4.32, we obtain the realized heritability of liability toward pyloric stenosis among males as

$$h^2 = \frac{2\mu'}{\mu_P} = \frac{2(T_P - T_O)}{\mu_P} = \frac{2(2.58 - 1.64)}{2.89} = 65\% \qquad (4.35)$$

Although Figure 4.14 depicts the underlying rationale for estimating the heritability of liability, the exact implementation is a little tedious. However, over a broad range of population incidences, the following approximation is sufficiently accurate for most purposes:

$$\log_{10}(B_R) \approx \left(\frac{-0.274h^2}{1 + 0.742h^2} \right) + \left(\frac{1 - 0.579h^2}{1 + 0.371h^2} \right) \log_{10}(B_G) \qquad (4.36)$$

Equation 4.36 generalizes the symbolism in Figure 4.14 in that B_G now represents the incidence of the trait in the general population (corresponds to B_P in Figure 4.14) and B_R represents the risk of the trait in first-degree relatives of affected individuals (corresponds to B_O in Figure 4.14). **First-degree relatives** share half their genes, and the most important first-degree relationships are parent-offspring and full siblings. Equation 4.36 is a satisfactory approximation over the broad range of values of B_G from 0.00001 to 0.20. Therefore, within this range, given any values of B_G and h^2 (or B_R and h^2), the corresponding value of B_R (or B_G) can easily be approximated from Equation 4.36. In the example of pyloric stenosis, $B_G = 0.005$ and $h^2 = 0.65$. Using Equation 4.36, $\log_{10}(B_R)$ is estimated as -1.276, hence $B_R \approx 0.05$. Equation 4.36 can also be adjusted for use with relatives other than those of the first degree. For the relationship between uncles/aunts and their nephews/nieces, replace h^2 by $h^2/2$, and for first cousins, replace h^2 by $h^2/4$.

Applications to Human Disease

Equation 4.36 implies that, for a fixed value of h^2, the relationship between $\log_{10}(B_R)$ and $\log_{10}(B_G)$ should be approximately linear. This is shown in Figure 4.15 along with the relationships expected for simple Mendelian dominant and simple Mendelian recessive inheritance. (The risk plotted for

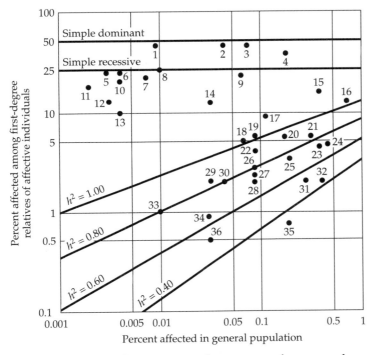

Figure 4.15 Risks of occurrence and recurrence of common abnormalities. The diagonal lines are the theoretical risks in first-degree relatives for threshold traits with the narrow-sense heritabilities indicated. The horizontal line for simple dominant indicates the risk in offspring or siblings of affected individuals, that for simple recessive indicates the risk in siblings of affected individuals. The traits are: (1) achondroplasia, (2) target-cell anemia, (3) periodic paralysis, (4) otosclerotic deafness, (5) retinoblastoma, (6) hemophilia, (7) albinism, (8) retinitis pigmentosum, (9) cystic fibrosis, (10) phenylketonuria, (11) osteogenesis imperfecta, (12) microphthalmos, (13) Hirschsprung disease, (14) deafmutism, (15) bipolar depressive disorder, (16) mental deficiency, (17) schizophrenia, (18) congenital dislocated hip, (19) multiple sclerosis, (20) strabismus, (21) pyloric stenosis, (22) anencephaly, (23) diabetes, (24) rheumatic fever, (25) spina bifida aperta, (26) clubfoot, (27) patent ductus, (28) cleft lip, (29) celiac disease, (30) cleft palate, (31) congenital heart disease, (32) epilepsy, (33) situs inversus viscerum, (34) exomphalos, (35) hydrocephaly, and (36) psoriasis. (Data from Newcombe 1964).

recessive inheritance refers to the full sibling of an affected individual, not the offspring of an affected individual.) Also plotted are observed data for many of the most common clinically relevant conditions. Note that the threshold traits, as a group, tend to be much more common than the simple Mendelian traits.

Linkage Analysis and Lod Scores

Major genetic risk factors for discrete traits can sometimes be sorted out by **linkage analysis.** In this approach, molecular markers scattered across the genome are tracked through pedigrees in hope of identifying a marker that is genetically linked to the risk factor. An example is shown in Figure 4.16, which depicts four pedigrees segregating for a trait determined by a dominant allele D. For simplicity we assume complete penetrance of the dominant allele. The symbols M and m represent alleles of a molecular marker, such as a RAPD, AFLP, or RFLP (Chapter 1), and for simplicity we assume that the linkage

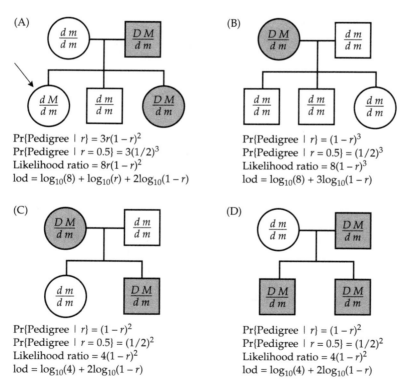

(A)

$\Pr\{\text{Pedigree} \mid r\} = 3r(1-r)^2$
$\Pr\{\text{Pedigree} \mid r = 0.5\} = 3(1/2)^3$
Likelihood ratio $= 8r(1-r)^2$
lod $= \log_{10}(8) + \log_{10}(r) + 2\log_{10}(1-r)$

(B)

$\Pr\{\text{Pedigree} \mid r\} = (1-r)^3$
$\Pr\{\text{Pedigree} \mid r = 0.5\} = (1/2)^3$
Likelihood ratio $= 8(1-r)^3$
lod $= \log_{10}(8) + 3\log_{10}(1-r)$

(C)

$\Pr\{\text{Pedigree} \mid r\} = (1-r)^2$
$\Pr\{\text{Pedigree} \mid r = 0.5\} = (1/2)^2$
Likelihood ratio $= 4(1-r)^2$
lod $= \log_{10}(4) + 2\log_{10}(1-r)$

(D)

$\Pr\{\text{Pedigree} \mid r\} = (1-r)^2$
$\Pr\{\text{Pedigree} \mid r = 0.5\} = (1/2)^2$
Likelihood ratio $= 4(1-r)^2$
lod $= \log_{10}(4) + 2\log_{10}(1-r)$

Figure 4.16 Pedigrees showing genetic linkage of a molecular marker M to a disease gene D.

phase of the double heterozygous parents is known to be $D\ M\ /\ d\ m$ in each pedigree. (This is called the **coupling** phase between D and M; the genotype $D\ m\ /\ d\ M$ is the **repulsion** phase.) The **frequency of recombination** between D and M is denoted r, and so a $D\ M\ /\ d\ m$ genotype is expected to produce gametes $D\ M$, $d\ m$, $D\ m$, and $d\ M$ in the frequencies $(1-r)/2$, $(1-r)/2$, $r/2$, and $r/2$, respectively. The first two are **nonrecombinant** types of gametes and the second two **recombinant** types.

Knowing the parental genotypes, we can write an expression for the probability of each observed pedigree in terms of the recombination fraction r using the binomial distribution. These probabilities are shown immediately below each pedigree. In this example there is only one recombinant progeny (indicated by the arrow). For each pedigree we can also write an expression for the probability of each pedigree under the assumption that $r = 0.5$ (independent segregation, which implies that the genes are either on different chromosomes or in one chromosome but widely separated). These probabilities are also given below each pedigree. For each pedigree the ratio of the probability of the pedigree given an arbitrary value of r, Pr{Pedigree$|r$}, to that with $r = 0.5$, Pr{Pedigree$|r = 0.5$}, is called the **likelihood ratio**. The **lod score** for each pedigree is the logarithm (base 10) of the likelihood ratio. (The name *lod* stands for *log-odds*, because the likelihood ratio is sometimes called the *odds*.)

Because the pedigrees are independent, the probabilities multiply across pedigrees, which mean that the likelihood ratios multiply also. Hence the lod scores are additive across pedigrees, because $\log_{10}(xy) = \log_{10}(x) + \log_{10}(y)$. For the pedigrees in Figure 4.16, therefore, we can add the lod scores to obtain the overall lod:

$$\text{lod} = \log_{10}(8 * 8 * 4 * 4) + \log_{10}(r) + 9\log_{10}(1-r)$$

The estimated value of r is the value that maximizes the lod score for all the pedigrees. The curve of lod against r is plotted in Figure 4.17. In this example the maximum lod is at $r = 0.10$, which makes intuitive sense because $1/10$ gametes is recombinant. In this case the evidence for linkage is not regarded as statistically significant because there are too few total progeny. In human genetics the convention is to regard a lod score greater than 3 as being statistically significant. The rationale is that the a priori odds against linkage in the human genome are about $50 : 1$ (modern genetic maps put it closer to $66 : 1$), and conventional statistical significance at the 5% level corresponds to odds of $20 : 1$. Hence, for pedigree data to override the a priori odds against linkage at a 5% level requires odds of $(50 \times 20) : (1 \times 1)$, or $1000 : 1$, which corresponds to a lod score of 3. On the other side, a lod score < -2 is generally regarded as significant evidence *against* linkage. Any value $-2 < \text{lod} < 3$ is considered uninformative in regard to linkage, and additional data are required.

Although 10 gametes is not sufficient to make a strong case for linkage in this example, 20 progeny would be sufficient. Suppose that the size of each

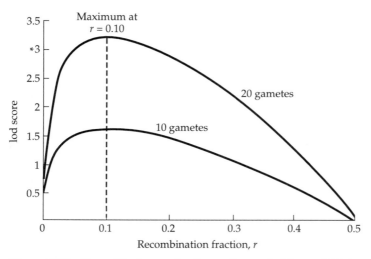

Figure 4.17   Overall lod score for the pedigrees in Figure 4.16 showing maximum lod at $r = 0.10$. The lod based on 10 gametes is not statistically significant; that based on 20 gametes is statistically significant (lod $\geq$ 3).

sibship in Figure 4.16 were doubled and that 2 of the total of 20 gametes were recombinant. Then you should try to convince yourself that the overall lod score in this case is

$$\text{lod} = \log_{10}(64 * 64 * 16 * 16) + 2\log_{10}(r) + 18\log_{10}(1 - r)$$

This is plotted in Figure 4.17 also, and it shows that the lod at $r = 0.10$ is greater than 3 and so is regarded as statistically significant evidence for linkage. A value of $r = 0.10$ is a frequency of recombination of 0.10 or a **percent recombination** of 10%. A standard unit of distance along a genetic map is the **centimorgan (cM),** which equals 1% recombination. Hence, in this case, the estimated map distance could also be stated as 10 cM.

Although the analysis in Figure 4.16 illustrates the underlying principles of linkage analysis, there are many complications in practice. For example, if the linkage phase is not known, the probability for each pedigree is a weighted average assuming an equal a priori probability of coupling and repulsion (Weir 1996). There are also problems with incomplete penetrance. For example, the major genetic risk factor for human breast cancer *BRCA1* has a penetrance of 37%, 66%, and 85% by ages 40, 55, and 80, so the ages of the women in the pedigrees must be taken into account; there is also genetic heterogeneity and environmental causes of breast cancer, yielding risks of 0.4%, 3%, and 8% for women of ages 40, 55, and 80 who do not carry *BRCA1* (Ford et al. 1994). Sorting out complexities such as these requires considerably more data than for simple Mendelian inheritance. For example, localizing a gene to 1 cM ($\approx 10^6$ bp) through comparisons of marker genotypes in pairs of affected siblings requires a median of 700 sib pairs for a gene that doubles the risk and 200 sib pairs for one that increases the risk by fivefold (Kruglyak and Lander 1995). Nontechnical discussion of alternative methods of dealing with such complexities can be found in Lander and Schork (1994) and Darvasi (1998).

QUANTITATIVE TRAIT LOCI (QTLs)

Any gene that influences a quantitative trait is called a **quantitative trait locus (QTL).** The opportunity to identify the chromosomal locations of QTLs, if not the genes themselves, has recently become possible owing to the prevalence of molecular markers and the relatively high density of such markers in the genetic maps of some species. In this section we consider some of the approaches for enumerating, genetic mapping, and identifying QTLs.

Number of Genes

When the number of genes affecting a quantitative trait is not large, and their effects are approximately equal, the number of QTLs can often be estimated from the means and variances observed in different strains and their hybrids and backcrosses. Suppose that a quantitative trait is affected by n equivalent, unlinked, and additive QTLs. For each locus the contribution to the phenotypic value is a, 0, and $-a$ for AA, AA', and $A'A'$ (hence the dominance $d = 0$). Consider two inbred lines, one homozygous for all n favorable alleles and having a phenotype of na and the other homozygous for all n unfavorable alleles and having a phenotype of $-na$. The total difference D between the inbred lines is therefore $D = na - (-na) = 2na$. Now suppose that the inbred lines are crossed to produce an F_1 generation. This generation is genetically homogeneous (heterozygous for all genes for which the parental inbred differ, homozygous otherwise), and its phenotypic variance should equal σ_e^2. The progeny of the F_1 generation constitutes the F_2 generation, and with no linkage its genotypic variance is given by Equation 4.8 as $\sigma_g^2 = 2pqa^2 = a^2/2$ because $p = q = 1/2$ (owing to the heterozygosity of all segregating genes in the F_1 generation). The total phenotypic variance in the F_2 generation is $\sigma_g^2 + \sigma_e^2$, but σ_g^2 can be estimated by subtracting out the F_1 variance as discussed in connection with Equation 4.4. Under these very simplified assumptions the number n of QTLs can be estimated from the **Wright-Castle index** (Castle 1921) as

$$n = \frac{D^2}{8\sigma_g^2} = \frac{(2na)^2}{8\left(\dfrac{na^2}{2}\right)} \tag{4.37}$$

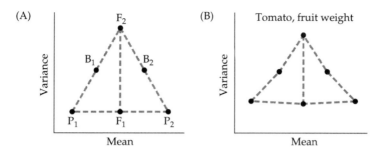

Figure 4.18 (A) Expected triangular relation between means and variances of phenotypic value among inbred parents (P), backcross progeny (B), and hybrid generations (F_1 and F_2) for an ideal quantitative trait determined by unlinked, equivalent, and completely additive genes. (B) Observed relation for an index of fruit weight in tomato. (After Lande 1981.)

When the simplifying assumptions are violated, application of the equation usually results in estimates of QTL number that are smaller than the actual number, and sometimes much smaller. Alternative methods of estimating the genetic variance using data from inbreds, hybrid, and backcrosses are discussed in Lande (1981), and biases in the estimate are discussed in Zeng (1992).

The scale in which phenotype is measured is important in using Equation 4.37 because the genes and alleles are assumed to be additive. In the ideal additive case, a plot of the means and variances of inbreds, hybrids, and backcrosses forms a triangle (Figure 4.18A). Therefore, in applying Equation 4.37, the phenotypes must be measured using a scale that yields an approximate triangle. With fruit weight in tomatoes, for example, an obvious scale of weight is in grams, but the scale that yields the triangle in Figure 4.18B is $x = \log_{10}[(\text{weight in grams}) - 0.153]$ (Lande 1981). Using this scale, $D = 1.552$ and $\sigma_g^2 = 0.0426$, from which Equation 4.37 yields $n = 7$ QTLs.

QTL Mapping

More than 50% of the phenotypic variance in tomato fruit weight can be accounted for by 6 QTLs that have been mapped genetically (Paterson et al. 1988). Genetic mapping of QTLs for a metric trait follows the same kind of rationale as that outlined in Figure 4.16 for a discrete trait, but for a metric trait each individual has its own phenotype measured on a continuous scale rather then being either affected or not affected.

Figure 4.19 shows the F_2 progeny expected from a cross of F_1 organisms of genotype $Q\,M\,/\,q\,m$, where Q and q are alleles affecting a metric trait and M and m are alleles of a molecular marker such as an RFLP for which heterozygotes can be identified. As usual, we denote the phenotypic contributions of QQ, Qq, and qq genotypes as a, d, and $-a$. The frequency of recombination between the QTL and the marker locus is denoted r with $0 \leq r \leq 0.5$. From the progeny genotypes in Figure 4.19, the following conditional probabilities may be deduced. Each symbol (for example, $\Pr\{QQ|MM\}$) means the probability that an individual has a particular QTL genotype (in this example, QQ), given that the individual is known to have a particular marker genotype (in this case, MM).

$$\Pr\{QQ|MM\} = (1-r)^2 \quad \Pr\{Qq|MM\} = 2r(1-r) \quad \Pr\{qq|MM\} = r^2$$

$$\Pr\{QQ|Mm\} = r(1-r) \quad \Pr\{Qq|Mm\} = r^2 + (1-r)^2 \quad \Pr\{qq|Mm\} = r(1-r)$$

$$\Pr\{QQ|mm\} = r^2 \quad \Pr\{Qq|mm\} = 2r(1-r) \quad \Pr\{qq|mm\} = (1-r)^2$$

From these conditional probabilities we can calculate the expected phenotype of each marker genotype based on the phenotypic values of the QTL genotypes. These expected values are

$$E(MM) = \Pr\{QQ|MM\}a + \Pr\{Qq|MM\}d + \Pr\{qq|MM\}(-a) = (1-2r)a + 2r(1-r)d$$

$$E(Mm) = \Pr\{QQ|Mm\}a + \Pr\{Qq|Mm\}d + \Pr\{qq|Mm\}(-a) = \left[r^2 + (1-r)^2\right]d$$

$$E(mm) = \Pr\{QQ|mm\}a + \Pr\{Qq|mm\}d + \Pr\{qq|mm\}(-a) = -(1-2r)a + 2r(1-r)d$$

$$(4.38)$$

Gametes and their frequencies	$Q\,M$ $\dfrac{1-r}{2}$	$q\,m$ $\dfrac{1-r}{2}$	$Q\,m$ $\dfrac{r}{2}$	$q\,M$ $\dfrac{r}{2}$
$Q\,M$ $\dfrac{1-r}{2}$	$QQ\,MM$ $\left(\dfrac{1-r}{2}\right)^2$	$Qq\,Mm$ $\left(\dfrac{1-r}{2}\right)^2$	$QQ\,Mm$ $\left(\dfrac{1-r}{2}\right)\left(\dfrac{r}{2}\right)$	$Qq\,MM$ $\left(\dfrac{1-r}{2}\right)\left(\dfrac{r}{2}\right)$
$q\,m$ $\dfrac{1-r}{2}$	$Qq\,Mm$ $\left(\dfrac{1-r}{2}\right)^2$	$qq\,mm$ $\left(\dfrac{1-r}{2}\right)^2$	$Qq\,mm$ $\left(\dfrac{1-r}{2}\right)\left(\dfrac{r}{2}\right)$	$qq\,Mm$ $\left(\dfrac{1-r}{2}\right)\left(\dfrac{r}{2}\right)$
$Q\,m$ $\dfrac{r}{2}$	$QQ\,Mm$ $\left(\dfrac{1-r}{2}\right)\left(\dfrac{r}{2}\right)$	$Qq\,mm$ $\left(\dfrac{1-r}{2}\right)\left(\dfrac{r}{2}\right)$	$QQ\,mm$ $\left(\dfrac{r}{2}\right)^2$	$Qq\,Mm$ $\left(\dfrac{r}{2}\right)^2$
$q\,M$ $\dfrac{r}{2}$	$Qq\,MM$ $\left(\dfrac{1-r}{2}\right)\left(\dfrac{r}{2}\right)$	$qq\,Mm$ $\left(\dfrac{1-r}{2}\right)\left(\dfrac{r}{2}\right)$	$Qq\,Mm$ $\left(\dfrac{r}{2}\right)^2$	$qq\,MM$ $\left(\dfrac{r}{2}\right)^2$

Figure 4.19 Progeny genotypes expected in the F_2 generation when a molecular marker M is genetically linked to a QTL with frequency of recombination r.

Because the genotype frequencies of *MM*, *Mm*, and *mm* in the F_2 generation are $1/4, 1/2$, and $1/4$, the mean phenotypic value of the F_2 generation can be verified to be $(1/4)E(MM) + (1/2)E(Mm) + (1/4)E(mm) = d/2$.

It should be intuitively clear that the marker genotype will have no association with the metric trait if it is unlinked to the QTL. There will also be no statistically significant association if the phenotypic effect of the QTL is too small. To see how linkage and QTL effects contribute jointly to any association, it is convenient to examine the regression coefficient of phenotype on the marker genotype. Generalizing the earlier discussion of the regression coefficient of offspring on parent in Equation 4.30, the regression coefficient of y on x equals the covariance between x and y divided by the variance of x. One type of regression is of the mean phenotype (P) against the number of M alleles in the marker genotype. The necessary genotypes, frequencies, and deviations are set out in Table 4.6. The covariance between phenotype and the number of M alleles equals the product of columns 1, 2, and 4, whereas the variance in the number of M alleles equals the product of column 1 and the squares of column 4. The regression coefficient of phenotype (P) on number of M alleles is therefore

$$b_{PM} = \frac{\text{Cov}(P, M)}{\text{Var}(M)} = \frac{(1/2)(1-2r)a}{(1/2)} = (1-2r)a \qquad (= a \text{ for } r = 0) \qquad (4.39)$$

The association between phenotype and number of M alleles will therefore be significant if $(1 - 2r)a$ is large enough, and this regression allows the estimation of a if r is known. Similarly we may regress phenotype (P) against whether or not the marker genotype is heterozygous. The relevant values are in the last two columns in Table 4.6. In this case, the covariance between phenotype and heterozygosity equals the product of columns 1, 2, and 6,

Table 4.6 Frequencies of a marker locus in the F_2 generation and phenotypic deviations due to a linked QTL

Genotype (frequency)	Mean phenotype (as deviation from population mean)	No. M alleles	No. M alleles as deviation	Heterozygosity	Heterozygosity as deviation
MM (1/4)	$(1-2r)a + 2r(1-r)d - d/2$	2	1	0	$-1/2$
Mm (1/2)	$[r^2 + (1-r)^2]d - d/2$	1	0	1	$1/2$
mm (1/4)	$-(1-2r)a + 2r(1-r)d - d/2$	0	-1	0	$-1/2$
Population mean	0	1	0	1/2	0

whereas the variance in heterozygosity equals the product of column 1 and the squares of column 6. The regression coefficient of phenotype (P) on heterozygosity is therefore

$$b_{PH} = \frac{\text{Cov}(P,H)}{\text{Var}(H)} = \frac{(1/4)(1-2r)^2 d}{(1/4)} = (1-2r)^2 d \qquad (= d \text{ for } r = 0) \qquad (4.40)$$

This regression of phenotype on heterozygosity depends only on the frequency of recombination and the dominance parameter d, and so d can be estimated if r is known.

Equations 4.39 and 4.40 are of special interest in the case $r = 0$, which means biologically that the QTL and the marker locus are one and the same gene, or at least are inseparable by recombination. When $r = 0$, the regression coefficient in Equation 4.39 estimates a, whereas the regression coefficient in Equation 4.40 estimates d. This special case brings us full circle to classical quantitative genetics, because a and d were originally defined through conceptual regressions of phenotype on number of favorable QTL alleles or heterozygosity (Fisher 1918; Falconer and Mackay 1996), long before the beginnings of modern molecular genetics.

For any QTL and marker locus that are associated, the values of a, d, and r could be estimated from the mean phenotypes of MM, Mm, and mm using Equation 4.38. This approach is somewhat precarious because there are three means from which to estimate three parameters and no opportunity for an independent test of goodness of fit. The preferred strategy is to have each QTL flanked by marker genes, which is called **interval mapping** (Lander and Botstein 1989). For the flanking markers there are nine genotypes that can be used to estimate a, d, and the frequencies of recombination r_1 and r_2 between the QTL and the flanking markers, with significance testing and estimation done by maximizing the lod scores of the likelihood ratios. For natural populations it is also necessary to estimate the allele frequencies of the QTL alleles. A more comprehensive approach is **composite interval mapping,** in which multiple regression is carried out on all the marker loci simultaneously (Zeng 1994; Jansen and Stam 1994). The analysis is carried out stepwise, identifying first the strongest effect and subtracting this out of the data, then identifying the next strongest effect, and so forth. Choosing appropriate lod scores for statistical significance is problematical because typically hundreds or thousands of comparisons are made, and in practice it is often preferred to permute the phenotypes randomly among all the genotypes so as to obtain an empirical significance level. For examples of this approach see Nuzhdin et al. (1997) and True et al. (1997).

Candidate Genes

QTL mapping has been used for a wide variety of traits in experimental organisms, including body weight, growth rate, obesity, atherosclerosis, and cancer susceptibility in the mouse, as well as hypertension, hyperactivity, and arthritis in the rat (Fisler and Warden 1997). It has also been widely used in pigs, poultry, cows, fish, and in many crop plants. Nevertheless, once a QTL has been assigned to a chromosomal subregion, identifying and isolating the gene remains a difficult problem. The reason is that the lod scores in QTL mapping typically have rather broad peaks as a function of chromosomal position, so additional data, and often a greater density of molecular markers in the relevant regions, are usually necessary to obtain greater precision in locating a QTL. Even then the isolation the QTL may be problematical, especially in a large genome. In the human genome 1 cM $\approx 10^6$ bp, which is a lot of DNA to characterize given the rather wide confidence limits (5–10 cM) typically accompanying a QTL localization.

Another approach to QTL identification is to use educated guesswork to identify **candidate genes** in advance based on their known functions. This approach may be especially promising in organisms targeted for genome sequencing, including *Arabidopsis thaliana* (mustard cress), *Caenorhabditis elegans* (nematode), *Drosophila melanogaster, Mus musculus,* and *Homo sapiens.* Even so, about half of all genes detected by genomic sequencing have no identified function, and many genes have two or more functions only one of which is known. Hence the candidate gene approach is unlikely to replace QTL mapping, but it serves as an important adjunct.

An example of the candidate gene approach is the identification of a QTL for anxiety as assessed by the neuroticism score on a standard personality test. One class of medications for anxiety and depression, including the widely prescribed Prozac®, selectively inhibits neuronal uptake of the neurotransmitter serotonin (5-hydroxytryptamine). Serotonin uptake is a normal process for helping to terminate neuronal stimulation caused by serotonin release. Hence the serotonin uptake transporter gene is an obvious candidate gene for traits related to anxiety or depression. The gene is *SLC6A4* in chromosome 17, and as it happens there is an insertion/deletion polymorphism of 44 bp in the promoter region approximately 1 kb upstream from the transcription initiation site. The insertion or "long" allele (*l*) has a frequency of about 57%, whereas the deletion or "short" allele (*s*) has a frequency of about 43%. Furthermore, in cells grown in culture, *l/l* cells have approximately 50 percent more mRNA for the transporter, and approximately 35 percent more membrane-bound transporter protein, than *s/l* or *s/s* cells.

An association between *SLC6A4* genotype and anxiety was observed in a study of 505 people genotyped for the transporter polymorphism and classified for personality traits from their responses to a set of questions (Lesch et al. 1996). Significant associations were found between the transporter genotype and the overall neuroticism score, and the highest correlations

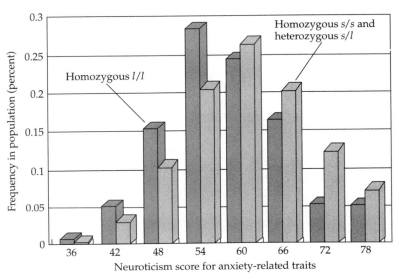

Figure 4.20 Distribution of neuroticism score for anxiety-related traits among genotypes for the serotonin uptake transporter. (Data from Lesch et al. 1996.)

were with the anxiety-related traits "tension" and "harm avoidance." A comparison between the genotypes with respect to neuroticism score is shown in Figure 4.20. Note that the *s* allele is dominant to *l* with respect to both gene expression and personality score.

Although there is a significant association, it is in the opposite direction than one might expect from the activity of selective serotonin uptake inhibitors. The effectiveness of these inhibitors might suggest that *s/–* genotypes should have reduced anxiety because they have reduced serotonin transporter, but the observed result is the other way around. There is also a great deal of overlap in the distributions: 51% of the *s/–* genotypes have a neuroticism score below the average for *l/l*, and 42% of the *l/l* genotypes score above the average for *s/–*. The *SLC6A4* polymorphism accounts for only about 8% of the phenotypic standard deviation in neuroticism score ($0.08\sigma_p$), and it is unlikely that current approaches for QTL mapping in a genetically heterogeneous population would have identified the *SLC6A4* polymorphism as being significantly associated with anxiety.

FURTHER READINGS

Bulmer, M. G. 1985. *The Mathematical Theory of Quantitative Genetics.* Clarendon Press, Oxford.

Comstock, R. 1996. *Quantitative Genetics with Special Reference to Plant and Animal Breeding.* Iowa State University Press, Ames.

Falconer, D. S. and T. Mackay. 1996. *Introduction to Quantitative Genetics.* Longman, Essex, UK.

Hartl, D. L. and A. C. Clark. 1997. *Principles of Population Genetics,* 3rd Ed. Sinauer Associates, Sunderland, MA.

Kempthorne, O. 1969. *An Introduction to Genetic Statistics.* Iowa State University Press, Ames.

Mather, K. and J. L. Jinks. 1982. *Biometrical Genetics,* 3rd Ed. Chapman and Hall, London.

Ott, J. *Analysis of Human Genetic Linkage,* 3rd Ed. Johns Hopkins Univ. Press, Baltimore, MD.

Pirchner, F. 1983. *Population Genetics in Animal Breeding.* Trans. by D. L. Frape. Plenum, New York.

Roff, D. A. 1997. *Evolutionary Quantitative Genetics.* Chapman and Hall, New York.

Weir, B. S., E. J. Eisen, M. M. Goodman and G. Namkoong (eds.). 1988. *Proceedings of the Second International Conference on Quantitative Genetics.* Sinauer Associates, Sunderland, MA.

Weiss, K. M. 1993. *Genetic Variation and Human Disease: Principles and Evolutionary Approaches.* Cambridge University Press, Cambridge.

Wright, S. 1968–1978. *Evolution and the Genetics of Populations.* Vol. 1, 1968: *Genetic and Biometric Foundations.* Vol. 2, 1969: *The Theory of Gene Frequencies.* Vol.3, 1977: *Experimental Results and Evolutionary Deductions.* Vol.4, 1978: *Variability within and among Natural Populations.* University of Chicago Press, Chicago.

PROBLEMS

4.1   A continuous trait is distributed approximately according to a normal distribution with mean 100 and standard deviation 15. What proportion of the population is expected to have a phenotypic value above 130? Below 85? Above 85?

4.2   A genetically heterogeneous population of wheat has a variance in the number of days to maturation of 40, whereas two inbred populations derived from it have a variance in the number of days to maturation of 10.

   a. What is the genotypic variance, σ_g^2, the environmental variance, σ_e^2, and the broad-sense heritability, H^2, of days to maturation in this population?

   b. If the inbred lines were crossed, what would be the predicted variance in days to maturation of the F_1 generation?

4.3   In comparing a quantitative trait in the F_1 and F_2 generations obtained by crossing two highly inbred strains:

   a. Which set of progeny provides an estimate of the environmental variance?

   b. What determines the variance of the other set of progeny?

4.4   In a cross between two cultivated inbred varieties of tobacco, the variance in leaf number per plant in the F_1 generation is 1.46 and in the F_2 generation it is 5.97. What are the genotypic and environmental variances? What is the broad-sense heritability in leaf number?

4.5   In Problem 2.7, the relative fitnesses of the *Standard* (*ST*) and *Arrowhead* (*AR*) inversions in *Drosophila pseudoobscura* were given as 0.47, 1.0, and 0.62, respectively, for *ST/ST*, *ST/AR*, and *AR/AR* genotypes. Let p represent the frequency of the *ST* inversion. Estimate the additive genetic variance in relative fitness due to these inversions for:

   a. $p = 0$ and $p = 1$
   b. $p = 0.2$ and $p = 0.8$
   c. $p = 0.38/(0.53 + 0.38)$
   d. Explain what is special about the previous value of p, and explain in words what the result implies about the relationship between the genotypic variance and the additive genetic variance.

4.6   The values of a and d depend on the scale of measurement of the trait. In Table 4.1, for example, different results are obtained when phenotypic values are measured in terms of x (the proportion of black coloration) rather than in terms of $\arcsin\sqrt{x}$. The appropriate scale of measurement for a trait is a scale in which the distribution of phenotypes is approximately normal, because normality is one of the underlying assumptions of the theory. Very often, data in proportions (x) are nonnormally distributed but become nearly normal when expressed as $\arcsin\sqrt{x}$. (Another "normalizing transformation" useful in many other examples is $\ln x$.) Wright (1968) provides the following data pertaining to the c^r and c^a (albino) alleles in guinea pigs.

	Degree of black coloration	
Genotype	**x = proportion black**	**$y = \arcsin\sqrt{x}$**
$c^r\, c^r$	0.87	68.87
$c^r\, c^a$	0.44	41.55
$c^a\, c^a$	0	0

Calculate a, d, and the degree of dominance (d/a) for:

   a. The proportions of black (x).
   b. For the transformed proportions $y = \arcsin\sqrt{x}$.

4.7   A meristic trait in a population is distributed as a binomial with individuals of phenotypic value i present in the proportion $[8!/(i!\,(8 - i!)](1/2)^8]$ for $i = 0, 1, 2, \ldots, 8$.

   a. What is the mean and variance of the distribution?
   b. If individuals with phenotypic values ≥ 6 are selected, what is the selection differential?
   c. If the narrow-sense heritability is 50%, what is the expected response to selection?

4.8   The prediction equation 4.15 is sometimes expressed as $R = i\sigma_p h^2$ where i is a new quantity called the "intensity of selection" that depends only on the proportion saved (B) and is used to compare different selection programs.

   a. Define i in terms of the selection differential S.
   b. Show that i depends only on B.

4.9   Below are data on the number i of abdominal bristles in samples from two consecutive generations G_1 and G_2 of an experiment in directional selection for increased bristle number in *Drosophila*. In the G_1 generation, flies with 22 or more bristles (enclosed in brackets) were mated together at random to form the G_2 generation. Estimate the realized heritability of the number of abdominal bristle in this experiment. (Data kindly provided by Trudy Mackay. In order to make the sexes comparable, the value of two has been added to the bristle number in males.)

i	G_1	G_2	i	G_1	G_2	i	G_1	G_2
15	0	2	20	20	13	25	[1]	3
16	21	4	21	12	14	26	0	2
17	5	7	22	[13]	12	27	0	0
18	18	16	23	[3]	6	28	0	2
19	17	17	24	[5]	3			

4.10   In analyzing data on oil content in maize, Lande (1981) found that values of log[(percent oil in kernels) + 1.87] approximated a triangular form like those in Figure 4.18. Using this scale of measurement, the means of two inbred lines were 0.513 and 1.122, respectively, and the phenotypic variances of the inbred lines, the F_1 generation, and the F_2 generation were 0.00142, 0.00053, 0.00030, and 0.00303, respectively. Use Equation 4.37 to estimate the Wright-Castle index corresponding to the number of genes under very simplified assumptions. For this calculation, make the estimate of the environmental variance in two different ways:

   a. As the mean of the variance in phenotypic value in the parental inbred lines
   b. As equal to the variance in phenotypic value in the F_1 generation.

4.11   Use Equations 4.13–4.15 to show that, if h^2 remains constant from generation to generation, the total response after n generations of selection is given by

$$\mu_n - \mu_0 = (S_0 + S_1 + \ldots + S_{n-1})h^2$$

where μ_0 and μ_n are the population means in generations 0 and n, respectively, and S_i is the selection differential applied in generation i. (The sum in this equation is called the cumulative selection differential.)

4.12   In the long-term experiment involving selection for high oil content in maize depicted in Figured 4.6, oil content increased linearly for 76 generations from an initial mean (μ_0) of 4.8% to a final mean (m_{76}) of 18.8%. In this same experiment, the cumulative selection differential increased by an approximately constant amount of 1.1 per generation (Dudley 1977). Estimate the realized heritability for the high-oil line in Figure 4.6.

4.13 The regression coefficient of offspring on the mean phenotypic value of the parents (often called the midparent value) can also be used to estimate the narrow-sense heritability. If b is the regression coefficient of offspring on midparent, then $h^2 = b$. (Note that there is no factor of $1/2$ in this case because both parents are involved.) Cook (1965) has studied shell breadth in 119 sibships of the snail *Arianta arbustorum*. For computational convenience, the data have been grouped into six categories.

Number of sibships	Midparent value (mm)	Offspring mean (mm)
22	16.25	17.73
31	18.75	19.15
48	21.25	20.73
11	23.75	22.84
4	26.25	23.75
3	28.75	25.42

Estimate the narrow-sense heritability of shell breadth from these data.

4.14 This problem introduces one form of R. A. Fisher's "fundamental theorem of natural selection." In a random mating population with alleles A and a at frequencies p and q, let the relative viabilities of AA, Aa, and aa be w_{AA}, w_{Aa}, and w_{aa}. The chain rule of differentiation implies that $d\bar{w}/dt = d\bar{w}/dp \times dp/dt$.

a. Use Equation 2.30 for $\bar{w}$ to obtain $d\bar{w}/dp$ and the numerator of Equation 2.31 as an approximation for dp/dt to find an expression for $d\bar{w}/dp$ in terms of the allele frequencies and the fitness parameters.

b. If the phenotypic values were written as a, d, and $-a$, respectively, how would one define a and d in terms of the fitness parameters?

c. Using these values, find a formula for the average effect of an allele substitution α in terms of the fitness parameters.

d. What is $d\bar{w}/dp$ expressed in terms of α? How is this expression related to the additive genetic variance in fitness?

4.15 In a population in HWE, show that the narrow-sense heritability of a trait completely determined by a single autosomal-recessive allele with frequency q equals $2q/(1 + q)$, which implies that $h^2 \approx 0$ for $q \approx 0$. (Hint: If $\sigma_e^2 = 0$, then $\sigma_g^2 = \sigma_a^2 + \sigma_d^2$.)

4.16 In a population in HWE, show that the heritability of a trait completely determined by a single autosomal dominant allele with frequency q equals $2(1 - q)/(2 - q)$, which implies that $h^2 \approx 1$ for $q \approx 0$.

4.17 Give an explanation in words of why $h^2 \approx 0$ for a rare recessive allele, whereas $h^2 \approx 1$ for a rare dominant allele, in terms of the correlation in phenotype between parents and offspring.

4.18 This problem demonstrates how the average effect of an allele substitution α can be defined in terms of regression. Suppose that a QTL with alleles A and A' at frequencies p and q is in HWE and that the QTL contributes a, d, or $-a$ units to phenotypic value in genotypes AA, AA', and $A'A'$, respectively. What is the regression coefficient of mean phenotypic value on number of A alleles?

4.19   Consider an X-linked QTL in a random mating population with alleles A and A' at frequencies p and q, respectively, in which the female genotypes AA, AA', and $A'A'$ have phenotypic values a, d, and $-a$, respectively, and male genotypes A and A' have phenotypic values a and $-a$, respectively. Show that the covariance in phenotypic value between mothers and their sons is given by $2p^2q^2a\alpha$, where α is the average effect of an allele substitution in females.

4.20   The "trait heritability" of a complex threshold trait is the conditional probability that an offspring is affected, given that one parent is affected, or $\Pr\{O|P\}$. The trait heritability is different, and usually smaller, than the heritability of liability. To express the trait heritability in terms of offspring-on-parent regression, code the phenotypic values such that "affected" = 1 and "not affected" = 0, which yields the following table:

Frequency in population	Parent phenotype	Offspring phenotype
p_1	1	1
p_2	1	0
p_3	0	1
p_4	0	0

where $p_1 + p_2 + p_3 + p_4 = 1$. Note that $p_1 + p_2 = p_1 + p_3 = b$, the overall frequency of affected persons in the population, and $p_1 = \Pr\{O|P\}\Pr\{P\} = \Pr\{O|P\}b$. Show that the covariance between parent and offspring equals $p_1 - b^2$ and that the variance among parents equals $b - b^2$. The regression coefficient of offspring on parent equals the ratio of these two quantities. Assuming $b^2 \ll b$, show that this ratio is approximately equal to $\Pr\{O|P\}$, which defines the trait heritability.

4.21   Renal stone disease is a threshold trait that occurs at a frequency of 0.4% in the general population and approximately 2.5% among the offspring of affected individuals. Solve Equation 4.36 to estimate the narrow-sense heritability of liability. Then use Equation 4.36 again to estimate the frequency of the trait in:

a. Brothers or sisters of affected individuals.
b. Nephews or nieces of affected individuals.
c. First cousins of affected individuals.

Solutions to the problems, worked out in full, can be found at the website www.sinauer.com/hartl/html

Literature Cited

Akashi, H. 1993. Synonymous codon usage in *Drosophila melanogaster*: Natural selection and translational accuracy. Genetics 136: 927–935.

Akashi, H. 1995. Inferring weak selection from patterns of polymorphism and divergence at "silent" sites in *Drosophila* DNA. Genetics 139: 1067–1076.

Akashi, H. 1997. Codon bias evolution in *Drosophila*: Population genetics of mutation-selection drift. Gene 205: 269–278.

Allison, A. C. 1964. Polymorphism and natural selection in human populations. Cold Spring Harbor Symp. Quant. Biol. 29: 139–149.

Anderson, W. W., C. Oshima, T. Watanabe, Th. Dobzhansky, and O. Pavlovsky. 1968. Genetics of natural populations. XXXIX. A test of the possible influence of two insecticides on the chromosomal polymorphisms in *Drosophila pseudoobscura*. Genetics 58: 423–434.

Ayala, F. J., B. S. W. Chang, and D. L. Hartl. 1993. Molecular evolution of the *Rh3* gene in *Drosophila*. Genetica 92: 23–32.

Ayala, F. J. and D. L. Hartl. 1993. Molecular drift of the *bride of sevenless (boss)* gene in *Drosophila*. Mol. Biol. Evol. 10: 1030–1040.

Barton, N. 1998. The geometry of adaptation. Nature 395: 751–752.

Barton, N. and M. Turelli. 1989. Evolutionary quantitative genetics: How little do we know? Annu. Rev. Genet. 23: 337–370.

Begun, D. J. and C. F. Aquadro. 1992. Levels of naturally occurring DNA polymorphism correlate with recombination rates in *D. melanogaster*. Nature 356: 519–520.

Bernardi, G. 1995. The human genome: Organization and evolutionary history. Annu. Rev. Genet. 29: 445–476.

Berry, A., J. Ajioka, and M. Kreitman. 1991. Lack of polymorphism on the *Drosophila* fourth chromosome resulting from selection. Genetics 129: 1111–1117.

Berry, A. J. and M. Kreitman. 1993. Molecular analysis of an allozyme cline: Alcohol dehydrogenase in *Drosophila melanogaster* on the East Coast of North America. Genetics 134: 869–893.

Bertranpetit, J., J. Sala, F. Calafell, P. A. Underhill, P. Moral, and D. Comas. 1995. Human mitochondrial DNA variation and the origin of Basques. Ann. Hum. Genet. 59: 63–81.

Blot, M., B. Hauer, and G. Monnet. 1994. The *Tn5* bleomycin resistance gene confers improved survival and growth advantage on *Escherichia coli*. Mol. General Genet. 242: 595–601.

Botstein, D., R. L. White, M. Skolnick, and R. W. Davis. 1980. Construction of a genetic linkage map in man using restriction length polymorphisms. Amer. J. Hum. Genet. 32: 314–331.

Braverman, J. M., R. R. Hudson, N. L. Kaplan, C. H. Langley, and W. Stephan. 1995. The hitchhiking effect on the site frequency spectrum of DNA polymorphisms. Genetics 140: 783–796.

Britten, R. J. 1997. Mobile elements inserted in the distant past have taken on important functions. Gene 205: 177–182.

Brookfield, J. F. Y. and P. M. Sharp. 1994. Neutralism and selectionism face up to DNA data. Trends Genet. 10: 109–111.

Brown, A. H. D. 1975. Sample sizes required to detect linkage disequilibrium between two or three loci. Theor. Pop. Biol. 8: 184–201.

Brown, W. M. 1980. Polymorphism in mitochondrial DNA of humans as revealed by restriction endonuclease analysis. Proc. Natl. Acad. Sci. USA 77: 3605–3609.

Bürger, R. and R. Lande. 1994. On the distribution of the mean and variance of a quantitative trait under mutation-selection-drift balance. Genetics 138: 901–912.

Buri, P. 1956. Gene frequency in small populations of mutant *Drosophila*. Evolution 10: 367–402.

Caballero, A. and P. D. Keightley. 1994. A pleiotropic nonadditive model of variation in quantitative traits. Genetics 138: 883–900.

Calafell, F. and J. Bertranpetit. 1994. Principal component analysis of gene frequencies and the origin of Basques. Amer. J. Phys. Anthro. 93: 201–215.

Capy, P., T. Langin, Y. Bigot, F. Brunet, M. J. Daboussi, G. Periquet, J. R. David, and D. L. Hartl. 1994. Horizontal transmission versus ancient origin: *mariner* in the witness box. Genetica 93: 161–170.

Carter, C. O. 1961. The inheritance of congenital pyloric stenosis. Brit. Med. Bull. 17: 251–254.

Castle, W. E. 1921. An improved method of estimating the number of genetic factors concerned in cases of blending inheritance. Science 54: 223.

Chakravarty, A. 1999. Population genetics—making sense out of sequence. Nature Genet. Suppl. 21: 56–60.

Charlesworth, B. 1985. The population genetics of transposable elements. pp. 213–232. In T. Ohta and K. Aoki (eds.), *Population Genetics and Molecular Evolution*. Springer-Verlag, Berlin.

Charlesworth, D. and B. Charlesworth. 1987. Inbreeding depression and its evolutionary consequences. Annu. Rev. Ecol. System. 18: 237–268.

Charlesworth, D. and B. Charlesworth. 1999. How was the *Sdic* gene fixed? Nature 400: 519–520.

Charlesworth, D., B. Charlesworth, and M. T. Morgan. 1995. The pattern of neutral molecular variation under the background selection model. Genetics 141: 1619–1632.

Charlesworth, B. and C. H. Langley. 1989. The population genetics of *Drosophila* transposable elements. Annu. Rev. Genet. 23: 251–287.

Charlesworth, B., P. Sniegowski, and W. Stephan. 1994. The evolutionary dynamics of repetitive DNA in eukaryotes. Nature 371: 215–220.

Clark, J. B., W. P. Maddison, and M. G. Kidwell. 1994. Phylogenetic analysis supports horizontal transmission of *P* transposable elements. Mol. Biol. Evol. 11: 40–50.

Clayton, G. A. and A. Robertson. 1955. Mutation and quantitative variation. Amer. Nat. 89: 151–158.

Cook, L. M. 1965. Inheritance of shell size in the snail *Arianta arbustorum*. Evolution. 19: 86–94.

Cross, S. R. H. and A. J. Birley. 1986. Restriction endonuclease map variation in the *Adh* region in populations of *Drosophila melanogaster*. Biochem. Genet. 24: 415–433.

Crow, J. F. and M. Kimura. 1970. *An Introduction to Population Genetics Theory*. Harper & Row, New York.

Crow, J. F. 1999. The odds of losing at genetic roulette. Nature 397: 293–294.

Curtsinger, J. W. 1984. Evolutionary landscapes for complex selection. Evolution 38: 359–367.

Curtsinger, J. W., P. M. Service, and T. Prout. 1994. Antagonistic pleiotropy, reversal of dominance, and genetic polymorphism. Amer. Nat. 144: 210–228.

Darvasi, A. 1998. Experimental strategies for the genetic dissection of complex traits in animal models. Nature Genet. 18: 19–24.

Darwin, C. 1859. *The Origin of Species*. John Murray, London.

David, J. R., H. Merçot, P. Capy, S. F. McEvey, and J. Herrewege. 1986. Alcohol tolerance and *Adh* gene frequencies in European and African populations of *Drosophila melanogaster*. Génétique, Sélection. Evolution 18: 405–416.

Dice, L. R. and W. B. Howard. 1951. Distance of dispersal by prairie deer mice from birthplace to breeding sites. Contr. Lab. Vert. Biol. Univ. Mich. 50: 1–15.

Dobzhansky, Th. and B. Spassky. 1963. Genetics of natural populations: XXXIV. Adaptive norm, genetic load, and genetic elite in *Drosophila pseudoobscura*. Genetics 48: 1467–1485.

Donnelly, P. and S. Tavare. 1995. Coalescents and genealogical structure under neutrality. Annu. Rev. Genet. 29: 401–421.

Drake, J. W. 1991. Spontaneous mutation. Annu. Rev. Genet. 25: 125–146.

Draye, X., P. Bullens, and F. A. Lints. 1994. Geographic variations of life history strategies in *Drosophila melanogaster*. 1. Analysis of wild-caught populations. Exper. Gerontol. 29: 205–222.

Dudley, J. W. 1977. 76 generations of selection for oil and protein percentage in maize. pp. 459–473. In E. Pollack, O. Kempthorne, and T. B. Bailey Jr. (eds.), *International Conference on Quantitative Genetics*. Iowa State University Press, Ames.

Dykhuizen, D. E., A. M. Dean, and D. L. Hartl. 1987. Metabolic flux and fitness. Genetics 115: 25–31.

Dykhuizen, D. and D. L. Hartl. 1980. Selective neutrality of 6PGD allozymes in *E. coli* and the effects of genetic background. Genetics 96: 801–817.

Eanes, W. F., M. Kirchner, and J. Yoon. 1993. Evidence for adaptive evolution of the *G6pd* gene in the *Drosophila melanogaster* and *Drosophila simulans* lineages. Proc. Natl. Acad. Sci. USA 90: 7475–7479.

Eanes, W. F., M. Kirchner, J. Yoon, C. H. Biermann, I. N. Wang, M. A. McCartney, and B. C. Verrelli. 1996. Historical selection, amino acid polymorphism and lineage-specific divergence at the *G6pd* locus in *Drosophila melanogaster* and *D. simulans*. Genetics 144: 1027–1041.

Easteal, S. 1985. The ecological genetics of introduced populations of the giant toad *Bufo marinus*. II. Effective population size. Genetics 110: 107–122.

Enfield, F. D. 1980. Long term effects of selection: The limits to response. pp. 69–86. In A. Robertson (ed.), *Selection Experiments in Laboratory and Domestic Animals*. Commonwealth Agricultural Bureaus, Slough, UK.

Engels, W. R. 1997. Invasions of *P* elements. Genetics 145: 11–15.

Epling, C. and T. Dobzhansky. 1942. Genetics of natural populations. VI. Microgeographic races in *Linanthus parryae*. Genetics 27: 317–332.

Ewens, W. J. 1972. The sampling theory of selectively neutral alleles. Theor. Popul. Biol. 3: 87–112.

Ewens, W. J. 1979. *Mathematical Population Genetics*. Springer-Verlag, New York.

Eyre-Walker, A. 1996. Synonymous codon bias is related to gene length in *Escherichia coli*: Selection for translational accuracy. Mol. Biol. Evol. 13: 864–872.

Eyre-Walker, A. and P. D. Keightley. 1999. High genomic deleterious mutation rates in hominids. Nature 397: 344–347.

Falconer, D. S. 1977. Some results of the Edinburgh selection experiments with mice. pp. 101–115. In E. Pollak, O. Kempthorne, and T. J. Bailey, Jr. (eds.), *International Conference on Quantitative Genetics*. Iowa State University Press, Ames.

Falconer, D. S. and T. F. C Mackay. 1996. *Introduction to Quantitative Genetics*, 4th Ed. Longman, Essex, UK.

Finnegan, D. J. and D. H. Fawcett. 1986. Transposable elements in *Drosophila*. Oxford Survey Eukaryotic Genet. 3: 1–62.

Fisher, R. A. 1918. The correlation between relatives on the supposition of Mendelian inheritance. Trans. Royal Soc. Edinburgh 52: 399–433.

Fisher, R. A. 1930. *The Genetical Theory of Natural Selection*. Oxford University Press, Oxford.

Fisler, J. S. and C. H. Warden. 1997. Mapping of mouse obesity genes: A generic approach to a complex trait. J. Nutrition 127: S1909–S1916.

Ford, D., D. F. Easton, D. T. Bishop, S. A. Narod, D. E. Goldgar, N. Haites, B. Milner, L. Allan, B. A. J. Ponder, J. Peto, S. Smith, M. Stratton, G. M. Lenoir, J. Feunteun, H. Lynch, A. Arason. R. Barkardottir, V. Egilsson, D. M. Black, D. Kelsell, N. Spurr, P. Devilee, C. J. Cornelisse, H. Varsen, J. M. Birch et al. 1994. Risks of cancer in *BRCA1*-mutation carriers. Lancet 343: 692–695.

Ford, E. B. and P. M. Sheppard. 1969. The *medionigra* polymorphism of *Panaxia dominula*. Heredity 24: 561–569.

Frankham, R. 1995. Inbreeding and extinction: A threshold effect. Conserv. Biol. 9: 792–799.

Fry, J. D., S. L. Heinsohn and T. F. C. Mackay. 1996. The contribution of new mutations to genotype-environment interaction for fitness in *Drosophila melanogaster*. Evolution 50: 2316–2327.

Fu, Y.-X. and W.-H. Li. 1993. Statistical tests of neutrality of mutations. Genetics 133: 693–709.

Fu, Y.-X. and W. H.-Li. 1997. Estimating the age of the common ancestor of a sample of DNA sequences. Mol. Biol. Evol. 14: 195–199.

Galton, F. 1889. *Natural Inheritance*. Macmillan, London.

Garcia-Fernàndez, J., J. R. Bayascas-Ramírez, G. Marfany, A. M. Muñoz-Mármol, A. Casali, J. Baguñà, and E. Saló. 1995. High copy number of highly similar *mariner*-like transposons in planarian (Platyhelminthe): Evidence for a trans-phyla horizontal transfer. Mol. Biol. Evol. 12: 421–431.

Gillespie, J. H. 1989. Lineage effects and the index of dispersion of molecular evolution. Mol. Biol. Evol. 6: 636–647.

Gillespie, J. H. and C. H. Langley. 1979. Are evolutionary rates really variable? J. Mol. Evol. 13: 27–34.

Golding, G. B., C. F. Aquadro, and C. H. Langley. 1986. Sequence evolution within populations under multiple types of mutation. Proc. Natl. Acad. Sci. USA 83: 427–431.

Hagemann, S., E. Haring, and W. Pinsker. 1996. Repeated horizontal transfer of *P* transposons between *Scaptomyza pallida* and *Drosophila bifasciata*. Genetica 98: 43–51.

Hardy, G. 1908. Mendelian proportions in a mixed population. Science 28: 49–50.

Harris, H. and D. A. Hopkinson. 1972. Average heterozygosity in man. Ann. Hum. Genet. 36: 9–20.

Hartl, D. L. and A. G. Clark. 1997. *Principles of Population Genetics*, 3rd Ed. Sinauer Associates, Sunderland MA.

Hartl, D. L. and D. E. Dykhuizen. 1981. Potential for selection among nearly neutral allozymes of 6-phosphogluconate dehydrogenase in *Escherichia coli*. Proc. Natl. Acad. Sci. USA 78: 6344–6348.

Hartl, D. L., D. E. Dykhuizen, and A. M. Dean. 1985. Limits of adaptation: The evolution of selective neutrality. Genetics 111: 655–674.

Hartl, D. L., A. R. Lohe, and E. R. Lozovskaya. 1997. Modern thoughts on an ancyent *marinere*: Function, evolution, regulation. Annu. Rev. Genet. 31: 337–358.

Hartl, D. L., E. R. Lozovskaya, D. I. Nurminsky, and A. R. Lohe,. 1997. What restricts the activity of *mariner*-like transposable elements? Trends Genet. 13: 197–201.

Hartl, D. L., E. N. Moriyama, and S. A. Sawyer. 1994. Selection intensity for codon bias. Genetics 138: 227–234.

Hartl, D. L. and S. A. Sawyer. 1988. Why do unrelated insertion sequences occur together in the genome of *Escherichia coli*? Genetics 118: 537–541.

Hartl, D. L. and C. H. Taubes,. 1996. Compensatory nearly neutral mutations: Selection without adaptation. J. Theor. Biol. 182: 303–309.

Hartl, D. L. and C. H. Taubes,. 1998. Towards a theory of evolutionary adaptation. Genetica 102/103: 525–533.

Haymer, D. S. and D. L. Hartl. 1983. The experimental assessment of fitness in *Drosophila*. II. A comparison of competitive and noncompetitive measures. Genetics 104: 343–352.

Hey, J. and R. M. Kliman. 1993. Population genetics and phylogenetics of DNA sequence variation at multiple loci within the *Drosophila melanogaster* species complex. Mol. Biol. Evol. 10: 804–822.

Hillis, D. M. and C. Moritz. 1990. *Molecular Systematics*. Sinauer Associates, Sunderland, MA.

Hillis, D. M., J. P. Huelsenbeck, and C. W. Cunningham. 1994. Application and accuracy of molecular phylogenies. Science 264: 671–677.

Hiraizumi, Y., L. Sandler, and J. F. Crow. 1960. Meiotic drive in natural populations of *Drosophila melanogaster*. III. Populational implications of the *segregation-distorter* locus. Evolution 14: 433–444.

Houle, D. 1998. How should we explain variation in the genetic variance of traits? Genetica 102/103: 241–253.

Hudson, R. R. 1990. Gene genealogies and the coalescent process. pp. 1–44. In D. Futuyma and J. Antonovics (eds.), *Oxford Surveys in Evolutionary Biology*, Vol. 7. Oxford University Press, New York.

Hudson, R. R. and N. L. Kaplan. 1995a. The coalescent process and background selection. Phil. Trans. Royal Soc. B 349: 19–23.

Hudson, R. R. and N. L. Kaplan. 1995b. Deleterious background selection with recombination. Genetics 141: 1605–1617.

Hudson, R. R., M. Kreitman, and M. Aguadé. 1987. A test of neutral molecular evolution based on nucleotide data. Genetics 116: 153–159.

Ikemura, T. 1985. Codon usage and tRNA content in unicellular and multicellular organisms. Mol. Biol. Evol. 2: 13–34.

Jansen, R. C. and P. Stam. 1994. High resolution of quantitative traits into multiple loci via interval mapping. Genetics 136: 1447–1455.

Jeffreys, A., J. V. Wilson, and S. L. Thein. 1985. Hypervariable 'minisatellite' regions in human DNA. Nature 314: 67–72.

Johannsen, W. 1903. Über Erblichkeit in Populationen und in reinen Linien. Gustav Fisher, Jena, Germany. [Translated, in part, pp. 21–26. In J. A. Peters (ed.), 1959, *Classic Papers in Genetics*, Prentice Hall, Englewood Cliffs, NJ.

Jukes, T. H. and C. R. Cantor. 1969. Evolution of protein molecules. pp. 21–132. In H. N. Munro (ed.), *Mammalian Protein Metabolism*. Academic Press, New York.

Kacser, H. and J. A. Burns. 1973. The control of flux. Symp. Soc. Exptl. Biol. 32: 65–104.

Karn, M. N. and L. S. Penrose. 1951. Birth weight and gestation time in relation to maternal age, parity, and infant survival. Annals Eugen. 16: 147–164.

Kerem, B., J. M. Rommens, J. A. Buchanan, D. Markiewicz, T. K. Cox, A. Chakravarti, M. Buchwald, and L.-C. Tsui. 1989. Identification of the cystic fibrosis gene: Genetic analysis. Science 245: 1073–1080.

Kettlewell, H. B. D. 1956. Further selection experiments on industrial melanism in the Lepidoptera. Heredity 10: 287–301.

Kimura, M. 1955. Solution of a process of random genetic drift with a continuous model. Proc. Natl. Acad. Sci. USA 41: 144–150.

Kimura, M. 1957. Some problems of stochastic processes in genetics. Ann. Math. Stat. 28: 882–901.

Kimura, M. 1964. Diffusion models in population genetics. J. Appl. Prob. 1: 177–232.

Kimura, M. 1968. Evolutionary rate at the molecular level. Nature 217: 624–626.

Kimura, M. 1980. A simple method for estimating evolutionary rates of base substitutions through comparative studies of nucleotide sequences. J. Mol. Evol. 16: 111–120.

Kimura, M. 1983. *The Neutral Theory of Molecular Evolution.* Cambridge University Press, Cambridge.

Kimura, M. and T. Ohta. 1971. *Theoretical Aspects of Population Genetics.* Princeton University Press, Princeton NJ.

Kliman, R. M. and J. Hey. 1993. DNA sequence variation at the *period* locus within and among species of the *Drosophila melanogaster* complex. Genetics 133: 375–387.

Kliman, R. M. and J. Hey. 1994. The effects of mutation and natural selection on codon bias in the genes of *Drosophila*. Genetics 137: 1049–1056.

Kolmogorov, A. 1931. Über die analytischen Methoden in der Wahrscheinlichkeitsrechnung. Math. Ann. 104: 415–458.

Kreitman, M. and R. R. Hudson. 1991. Inferring the evolutionary histories of the *Adh* and *Adh-dup* loci in *Drosophila melanogaster* from patterns of polymorphism and divergence. Genetics 127: 565–582.

Kruglyak, L. and E. S. Lander. 1995. High-resolution genetic mapping of complex traits. Amer. J. Hum. Genet. 56: 1212–1223.

Kruse, H. and H. Sorum. 1994. Transfer of multiple drug resistance plasmids between bacteria of diverse origins in natural microenvironments. Appl. Environ. Microbiol. 60: 4015–4021.

Lande, R. 1981. The minimum number of genes contributing to quantitative variation between and within populations. Genetics 99: 541–553.

Lander, E. S. and D. Botstein. 1989. Mapping Mendelian factors underlying quantitative traits using RFLP linkage maps. Genetics 121: 185–199.

Lander, E. S. and N. J. Schork. 1994. Genetic dissection of complex traits. Science 265: 2037–2048.

Langley, C. H., J. F. Y. Brookfield, and N. Kaplan. 1983. Transposable elements in Mendelian populations. I. A theory. Genetics 104: 457–471.

Leroi, A. M., S. B. Kim, and M. R. Rose. 1994. The evolution of phenotypic life-history trade-offs: An experimental study using *Drosophila melanogaster*. Amer. Nat. 144: 661–676.

Lesch, K.-P., D. Bengel, A. Heils, S. Z. Sabol, B. D. Greenberg, S. Petri, J. Benjamin, C. R. Müller, D. H. Hamer, and D. L. Murphy. 1996. Association of anxiety-related traits with a polymorphism in the serotonin transporter gene regulatory region. Science 274: 1527–1531.

Lewontin, R. C. 1974. *The Genetic Basis of Evolutionary Change.* Columbia University Press, New York.

Lewontin, R. C. 1991. Electrophoresis in the development of evolutionary genetics: Milestone or millstone? Genetics 128: 657–662.

Li, W.-H. 1977. Distribution of nucleotide differences between two randomly chosen cistrons in a finite population. Genetics 85: 331–337.

Li, W.-H. 1997. *Molecular Evolution.* Sinauer Associates, Sunderland, MA.

Li, W.-H., C.-I. Wu, and C.-C. Luo. 1985. A new method for estimating synonymous and nonsynonymous rates of nucleotide substitution. Mol. Biol. Evol. 2: 150–174.

Lohe, A. R., E. N. Moriyama, D.-A. Lidholm, and D. L. Hartl. 1995. Horizontal transmission, vertical inactivation, and stochastic loss of *mariner*-like transposable elements. Mol. Biol. Evol. 12: 62–72.

Long, M. and C. H. Langley. 1993. Natural selection and origin of *jingwei*, a chimeric processed functional gene in *Drosophila*. Science 260: 91–95.

Lucotte, G. and G. Mercier. 1998. Distribution of *CCR5*. gene 32–bp deletion in Europe. J. Acquired Immune Deficiency Syndromes and Human Retrovirology. 19: 174–177.

Lynch, M. 1988. The rate of polygenic mutation. Genet. Res. 51: 137–148.

Lynch, M. and W. G. Hill. 1986. Phenotypic evolution by neutral mutation. Evolution 40: 915–935.

Mackay, T. F. C., J. D. Fry, R. F. Lyman, and S. V. Nuzhdin. 1994. Polygenic mutation in *Drosophila melanogaster:* Estimates from response to selection of inbred strains. Genetics 136: 937–951.

May, R. M. 1985. Evolution of pesticide resistance. Nature 315: 12–13.

McDonald, J. H. and M. Kreitman. 1991. Adaptive protein evolution at the *Adh* locus in *Drosophila*. Nature 351: 652–654.

Miller, W. J., J. F. McDonald and W. Pinsker. 1997. Molecular domestication of mobile elements. Genetica 100: 261–270.

Mukai, T., T. K. Watanabe and O. Yamaguchi. 1974. The genetic structure of natural populations of *Drosophila melanogaster*. XII. Linkage disequilibrium in a large local population. Genetics 77: 771–793.

Munte, A., M. Aguadé, and C. Segarra. 1997. Divergence of the *yellow* gene between *Drosophila melanogaster* and *D. subobscura:* Recombination rate, codon bias and synonymous substitutions. Genetics 147: 165–175.

National Research Council. 1992. *DNA Technology in Forensic Science.* National Academy Press, Washington, DC.

National Research Council. 1996. *The Evaluation of Forensic DNA Evidence.* National Academy Press, Washington, DC.

Neal, N. P. 1935. The decrease in yielding capacity in advanced generations of hybrid corn. J. Amer. Soc. Agron. 27: 666–670.

Nei, M. 1996. Phylogenetic analysis in molecular evolutionary genetics. Annu. Rev. Genet. 30: 371–403.

Nevo, E. 1978. Genetic variation in natural populations: Patterns and theory. Theoret. Pop. Biol. 13: 121–177.

Newcombe, H. B. 1964. In M. Fishbein (ed.), *Papers and Discussions of the Second International Conference on Congenital Malformations.* The International Medical Congress, New York.

Nikitin, A. G. and R. C. Woodruff. 1995. Somatic movement of the *mariner* transposable element and lifespan of *Drosophila* species. Mutation Research 338: 43–49.

Novick, A. 1955. Mutagens and antimutagens. Brookhaven Symp. Biol. 8: 201–215.

Nunney, L. 1996. The response to selection for fast larval development in *Drosophila melanogaster* and its effect on adult weight: An example of a fitness trade-off. Evolution 50: 1193–1204.

Nurminsky, D. I., M. V. Nurminskaya, D. De Aguiar, and D. L. Hartl. 1998. Selective sweep of a newly evolved sperm-specific gene in *Drosophila*. Nature 396: 572–575.

Nurminsky, D. I. and D. L. Hartl. 1999. How was the *Sdic* gene fixed? Nature 400: 520.

Nuzhdin, S. V., E. G. Pasyukova, C. L. Dilda, Z.-B. Zeng, and T. F. C. Mackay. 1997. Sex-specific quantitative trait loci affecting longevity in *Drosophila melanogaster*. Proc. Nat. Acad. Sci. USA 94: 9734–9739.

Oakeshott, J. G., J. B. Gibson, P. R. Anderson, W. R. Knibb, D. G. Anderson, and G. K. Chambers. 1982. Alcohol dehydrogenase and glycerol-3-phosphate dehyrdogenase clines in *Drosophila melanogaster* in different continents. Evolution 39: 86–96.

Oakeshott, J. G., G. K. Chambers, J. B. Gibson, W. F. Eanes, and D. A. Willcocks. 1983. Geographic variation in *G6pd* and *Pgd* allele frequencies in *Drosophila melanogaster*. Heredity 50: 67–72.

O'Brien, S. J., D. E. Wildt, M. Bush, T. M. Caro, C. FitzGibbon, I. Aggundey, and R. E. Leakey. 1987. East African cheetahs: Evidence for two population bottlenecks. Proc. Natl. Acad. Sci. USA 84: 508–511.

Ohta, T. 1992. The nearly neutral theory of molecular evolution. Annu. Rev. Ecol. Syst. 23: 263–286.

Ohta, T. 1973. Slightly deleterious mutant substitutions in evolution. Nature 246: 96–98.

Olson, M. V., L. Hood, C. Cantor, and D. Botstein. 1989. A common language for physical mapping of the human genome. Science 245: 1434–1440.

Orr, H. A. 1998. The population genetics of adaptation: The distribution of factors fixed during adaptive evolution. Evolution 52: 935–949.

Parsch, J., W. Stephan, and S. Tanda. 1998. Long-range base pairing in *Drosophila* and human mRNA sequences. Mol. Biol. Evol. 15: 820–826.

Paterson, A. H., E. S. Lander, J. D. Hewitt, S. Peterson, S. E. Lincoln, and S. D. Tanksley. 1988. Resolution of quantitative traits into Mendelian factors by using a complete linkage map of restriction fragment length polymorphisms. Nature 335: 721–726.

Penny, D., E. E. Watson, R. E. Hickson, and P. J. Lockhart. 1993. Some recent progress with methods for evolutionary trees. New Zealand J. Botany 31: 275–288.

Patterson, C., D. M. Williams, and C. J. Humphries. 1993. Congruence between molecular and morphological phylogenies. Annu. Rev. Ecol. System 24: 153–188.

Perlitz, M. and W. Stephan. 1997. The mean and variance of the number of segregating sites since the last hitchhiking event. J. Math. Biol. 36: 1–23.

Petrov, D. A., E. R. Lozovskaya, and D. L. Hartl. 1996. High intrinsic rate of DNA loss in *Drosophila*. Nature 384: 346–349.

Petrov, D. A., T. Sangster, J. S. Johnston, D. L. Hartl, and K. L. Shaw,. 1999. Evidence for DNA loss as a determinant of genome size. Unpublished data.

Pier, G. B., M. Grout, T. Zaidi, G. Meluleni, S. S. Mueschenborn, G. Banting, R. Ratcliff, M. J. Evans, and W. H. Colledge. 1998. Salmonella typhi used CFTR to enter intestinal epithelial cells. Nature 393: 79–82.

Pirchner, F. 1969. *Population Genetics in Animal Breeding*. W. H. Freeman, San Francisco.

Reed, T. E. and J. V. Neel. 1959. Huntington's chorea in Michigan. Amer. J. Hum. Genet. 11: 107–136.

Robertson, F. W. 1957. Studies in quantitative inheritance. XI. Genetic and environmental correlation between body size and egg production in *Drosophila melanogaster*. J. Genet. 55: 428–443.

Robertson, H. M. 1993. The *mariner* transposable element is widespread in insects. Nature 362: 241–245.

Robertson, H. M. and D. J. Lampe. 1995. Recent horizontal transfer of a *mariner* transposable element among and between Diptera and Neuroptera. Mol. Biol. Evol. 12: 850–862.

Robertson, H. M. and E. G. MacLeod. 1993. Five major subfamilies of *mariner* transposable elements in insects, including the Mediterranean fruit fly, and related arthropods. Insect Mol. Biol. 2: 125–139.

Robinson, H. F., R. E. Comstock, and P. H. Harvey. 1949. Estimates of heritability and degree of dominance in corn. Agron. J. 41: 353–359.

Russell, W. A. 1974. Comparative performance for maize hybrids representing different eras of maize breeding. Annu. Corn and Sorghum Res. Conf. 29: 81–101.

Sawyer, S. A., D. E. Dykhuizen, and D. L. Hartl. 1987. A confidence interval for the number of selectively neutral amino acid polymorphisms. Proc. Natl. Acad. Sci. USA 84: 6225–6228.

Sawyer, S. A., D. E. Dykhuizen, R. F. DuBose, L. Green, T. Mutangadura-Mhlanga, D. F. Wolczyk, and D. L. Hartl. 1987. Distribution and abundance of insertion sequences among natural isolates of *Escherichia coli*. Genetics 115: 51–63.

Sawyer, S. A. and D. L. Hartl. 1986. Distribution of transposable elements in prokaryotes. Theoret. Pop. Biol. 30: 1–17.

Sawyer, S. A. and D. L. Hartl. 1992. Population genetics of polymorphism and divergence. Genetics 132: 1161–1176.

Selander, R. K. and S. Y. Yang. 1969. Protein polymorphism and genic heterozygosity in a wild population of the house mouse (*Mus musculus*). Genetics 63: 653–667.

Sharp, P. M. and W.-H. Li. 1987. The codon adaptation index: A measure of directional codon usage bias and its potential application. Nucleic Acids Res. 15: 1281–1295.

Shields, D. C., P. M. Sharp, D. G. Higgins, and F. Wright. 1988. "Silent" sites in *Drosophila* are not neutral: Evidence of selection among synonymous codons. Mol. Biol. Evol. 5: 704–716.

Simmons, M. F. and J. F. Crow. 1977. Mutations affecting fitness in *Drosophila* populations. Annu. Rev. Genet. 11: 49–78.

Slatkin, M. 1985. Rare alleles as indicators of gene flow. Evolution 39: 53–65.

Slatkin, M. 1996. Gene genealogies within mutant allelic classes. Genetics 143: 579–587.

Smith, C. 1975. Quantitative inheritance. pp. 382–441. In G. Fraser and O. Mayo (eds.), *Textbook of Human Genetics*. Blackwell, Oxford.

Sprague, G. F. 1978. Introductory remarks to the session on the history of hybrid corn. Pp. 11–12. *in* D. B. Walden (ed.) *Maize Breeding and Genetics*. Wiley, New York.

Stam, L. F. and C. C. Laurie. 1996. Molecular dissection of a major gene effect on a quantitative trait: The level of alcohol dehydrogenase expression in *Drosophila melanogaster*. Genetics 144: 1559–1564.

Stocker, B. A. D. 1949. Measurements of rate of mutation of flagellar antigenic phase in *Salmonella typhimurium*. J. Hygiene 47: 398–412.

Sved, J. A. 1975. Fitness of third chromosome homozygotes in *Drosophila melanogaster*. Genet. Res. 25: 197–200.

Tajima, F. 1983. Evolutionary relationship of DNA sequences in finite populations. Genetics 105: 437–460.

Tajima, F. 1989. Statistical method for testing the neutral mutation hypothesis by DNA polymorphism. Genetics 123: 585–595.

Teissier, G. 1942. Persistence d'un gène léthal dans une population de *Drosophiles*. Compt. Rend. Acad. Sci. 214: 327–330.

True, J. R., J. J. Liu, L. F. Stam, Z.-B. Zeng, and C. C. Laurie. 197. Quantitative genetic analysis of divergence in male secondary sexual traits between *Drosophila simulans* and *Drosophila mauritiana*. Evolution 51: 816–832.

Turelli, M. and N. Barton. 1990. Dynamics of polygenic characters under selection. Theor. Pop. Bio. 38: 1–57.

Turelli, M. and N. H. Barton. 1994. Genetic and statistical analysis of strong selection on polygenic traits: What, me normal? Genetics 138: 913–941.

Valdes, A. M. and G. Thomson. 1997. Detecting disease-predisposing variants: The haplotype method. Amer. J. Hum. Genet. 60: 703–716.

van Delden, W., A. C. Boerma, and A. Kamping. 1978. The alcohol dehyrdogenase polymorphism in populations of *Drosophila melanogaster*. I. Selection in different environments. Experientia 31: 418–420.

Vilá, C. P. Savolainen, J. E. Maldonado, I. R. Amorim, J. E. Rice, R. L. Honeycutt, K. A. Crandall, J. Lundeberg, and R. K. Wayne. 1997. Multiple and ancient origins of the domestic dog. Science 276: 1667–1689.

Vos, P., R. Hogers, M. Bleeker, M. Reijans, T. van de Lee, M. Hornes, A. Frijters, J. Pot, J.

Peleman, M. Kuiper, and M. Zabeau. 1995. AFLP: A new technique for DNA fingerprinting. Nucleic Acids Res. 23: 4407–4414.

Watterson, G, A. 1975. On the number of segregating sites in genetic models without recombination. Theor. Popul. Biol. 7: 256–276.

Watterson, G. A. 1985. Estimating species divergence times using multilocus data. pp. 1–44. In T. Ohta and K. Aoki (eds.), *Population Genetics and Molecular Evolution.* Springer-Verlag, Berlin.

Weinberg, W. 1908. On the demonstration of heredity in man. Naturkunde in Wurttemberg, Stuttgart. 64: 368–382. [Original in German. Translated in S. H. Boyer IV (ed.), 1963, *Papers on Human Genetics,* Prentice-Hall, Englewood Cliffs, NJ.]

Weir, B. S. 1996. *Genetic Data Analysis II.* Sinauer Associates, Sunderland, MA.

Welsh, J. and M. McClelland. 1990. Fingerprinting genomes using PCR with arbitrary primers. Nucleic Acids Res. 18: 7213–7318.

Whittam, T. S. H. Ochman, and R. K. Selander. 1983. Multilocus genetic structure in natural populations of *Escherichia coli.* Proc. Natl. Acad. Sci. USA 80: 1751–1755.

Williams, J. G. K., A. R. Kubelik, K. J. Livak, J. A. Rafalski, and S. V. Tingery. 1990. DNA polymorphisms amplified by arbitrary primers are useful as genetic markers. Nucleic Acids Res. 18: 7213–7318.

Wright, F. 1990. The 'effective number of codons' used in a gene. Gene 87: 23–29.

Wright, S. 1934. Physiological and evolutionary theories of dominance. Amer. Nat. 68: 25–53.

Wright, S. 1938. The distribution of gene frequencies in populations of polyploids. Proc. Natl. Acad. Sci. USA 24: 372–377.

Wright, S. 1945. The differential equation of the distribution of allele frequencies. Proc. Natl. Acad. Sci. USA 31: 382–389.

Wright, S. 1968. *Evolution and the Genetics of Populations.* Vol. 1. *Genetic and Biometric Foundations.* University of Chicago Press, Chicago.

Wright, S. 1977. *Evolution and the Genetics of Populations.* Vol. 3. *Experimental results and Evolutionary Deductions.* University of Chicago Press, Chicago.

Wright, S. 1978. *Evolution and the Genetics of Populations.* Vol. 4. *Variability Within and Among Natural Populations.* University of Chicago Press, Chicago.

Yang, Z. 1994. Statistical properties of the maximum likelihood method of phylogenetic estimation and comparison with distance matrix methods. Systematic Biol. 43: 329–342.

Yang, Z. 1996. Phylogenetic analysis using parsimony and likelihood methods. J. Mol. Evol. 42: 294–307.

Yang, Z., R. Nielsen, and M. Hasegawa. 1998. Models of amino acid substitution and applications to mitochondrial protein evolution. Mol. Biol. Evol. 15: 1600–1611.

Zapata, C. and G. Alvarez. 1993. On the detection of nonrandom associations between DNA polymorphisms in natural populations of *Drosophila.* Mol. Biol. Evol. 10: 823–841.

Zeng, L. W., J. M. Comeron, B. Chen, and M. Kreitman. 1998. The molecular clock revisited: The rate of synonymous vs. replacement change in *Drosophila.* Genetica 103: 369–382.

Zeng, Z.-B. 1992. Correcting the bias of Wright's estimates of the number of genes affecting a quantitative character: A further improved method. Genetics 131: 987–1001.

Zeng, Z.-B. 1994. Precision mapping of quantitative trait loci. Genetics 136: 1457–1468.

Zharkikh, A. 1994. Estimation of evolutionary distances between nucleotide sequences. J. Mol. Evol. 39: 315–329.

Index

ABOUT THE BOOK

Editor: Andrew D. Sinauer
Project Editor: Carol J. Wigg
Copy Editor: Roberta Lewis
Production Manager: Christopher Small
Book Design: Michele Ruschhaupt
Cover Design: Wendy Beck
Book Production in QuarkXPress: Michele Rauchhaupt
Cover Manufacture: Henry N. Sawyer Company, Inc.
Book Manufacture: Courier Companies, Inc.